Health for the Farmer

Health for the Farmer

C. F. Stanford

FARMING PRESS

First published 1991

A catalogue record for this book is available
from the British Library

ISBN 0 85236 221 8

Published by Farming Press Books
4 Friars Courtyard, 30–32 Princes Street
Ipswich IP1 1RJ, United Kingdom

Distributed in North America
by Diamond Farm Enterprises,
Box 537, Alexandria Bay, NY 13607, USA

Cover design by Andrew Thistlethwaite
Photo by CNRI/Science Photo Library
Typeset by Galleon Photosetting, Ipswich
Printed and bound in Great Britain by Butler & Tanner, Frome, Somerset

Contents

Acknowledgements

I would like to thank Dr Gerald Hall for his encouragement to investigate industrial diseases in Northern Ireland and the many colleagues with whom I have worked in pursuit of this. Special thanks are due to Professor D. I. H. Simpson, Doctors J. H. Connolly, P. V. Coyle, W. A. Ellis, W. I. Montgomery, E. T. M. Smyth and Mr J. Evans, Mrs Barbara Martin and Mrs Caroline Simpson.

Funding for my research into farmers' health was received from EMAS, Northern Ireland Health and Safety Committee, Northern Ireland Milk Marketing Board, Northern Ireland Livestock Commissions and the Northern Ireland Farmer's Union.

Earlier studies of tropical medicine in the Indian subcontinent and Egypt were funded by the Nuffield Foundation.

None of the work would have been possible without the co-operation of all those farmers in Northern Ireland who volunteered for the studies. Special thanks are due to Arthur Hanna, William Brown, William McIlroy, John Boyd and Cecil Gowdy who, by their humanity and hard work over the years, stimulated my interest in farming.

The manuscript was prepared by Mrs Janice Gibney, Miss Gillian Wilson and Lucy Stanford. The overall project was co-ordinated by Mrs Julanne Arnold of Farming Press. This book would not have been possible without their help.

Finally, I would like to thank my wife Rosemary for her patience and encouragement throughout the period of both research and the writing of this book.

CFS

Foreword

Quite rightly, increasing attention is being paid to health and safety issues in the workplace both by national governments and international bodies.

Education and enforcement of regulations are most difficult to achieve in small work units. Nowhere more is this the case than in farming, which may have the added disadvantage of isolation. A further hazard is the involvement of young children and the elderly.

The author of the book has taken a personal interest in the health and safety of farmers and their families and has carried out valuable research among farmers in Northern Ireland. I would therefore strongly recommend the book to all those concerned with reducing accidents and ill-health on the farm.

DR J. G. HALL FFOM
Senior Employment Medical Adviser
Employment Medical Advisory Service (NI)

To the late David Allen,
who loved the land and reaped its harvests

Introduction

This book has been written to give the reader a taste for investigating health matters. It describes some of the factors that might contribute to both the health and ill-health of farmers and their families and it is clear that they are exposed to many more risks than they or the community at large realise. It is hoped that the information will enable farmers to understand the risks better and to recognise the symptoms and signs of diseases in both themselves and their animals. Awareness of these illnesses at an early stage is important if serious complications are to be avoided, and in most countries there are plenty of professionals available to guide and help. However, it is even more important to try to prevent these diseases from occurring in the first place. For this to happen, a major change will be required in the outlook of the farming community.

> **Prevention of disease is an attitude of mind.**

Farmers must become aware of the diseases that are prevalent in their own community. This applies to both human and animal health. Public health authorities and doctors can supply the human health statistics, and government and agricultural departments and veterinary surgeons can give relevant information about animals. Some countries have health and safety committees and these provide invaluable advice, especially about accident prevention and the dangers from chemicals.

Professional groups are now agreed that the only way to reduce the overwhelming burden of disease in the world population is to make people better educated about the risks. Ask questions and

1

persist until you get the answers. Farmers are in danger of being bypassed by this explosion of knowledge. A book of this size can only hope to summarise the problems that face in farmers and their families across the world.

Farmers cannot act in isolation and expect dramatic results. Governments can only react to the problems if they have correct information for their statistics. All farming accidents, no matter how small, must be reported. Check with your doctor if an illness you or your family is suffering from is a recognised occupational illness eligible for compensation (Appendix 1) or one that must be notified to the public health authority (Appendix 2).

It might be worth asking your Member of Parliament or state legislature if there is a government body with special responsibility for the health of farmers, as there is for the armed services, the miners and most large industries. The provision of such a service for an individual contractor or a self-employed person may be more difficult to achieve, but it is not impossible.

The population rightly expects to be able to buy good food at a reasonable price. It is reasonable for farmers to expect that price to include the cost of health care.

1

The Farmer as a Member of the Community

It is generally assumed that farmers are healthier than the rest of the society in which they live. After all, they work outdoors in 'good, clean, unpolluted air' and as the producers of food they can select and eat the best produce. Likewise, the exercise they take should help to keep weight down, with all the social and medical benefits that this brings, including a reduction in coronary artery and other arterial diseases.

There is little doubt from the available statistics that some diseases are in fact less prevalent in farmers, for instance cancer of the lung. This low incidence is usually thought to be due to lower exposure to atmospheric carcinogens in the countryside than in an urban or industrial setting. Some of the other concepts about healthy farmers do, however, need to be challenged.

Human growth takes place up until the age of eighteen. Farmers tend to be 1–2 cms taller than similarly aged industrial workers. This may be due to genetic influences—those families which were stronger and taller tending to remain in farming while others left for less strenuous work. It might also be due to better or more plentiful food during the formative years giving them a growth advantage over their urban contemporaries. The fact that young men from all over the UK who are applying to become soldiers are gradually becoming taller would suggest that better food and less infection in infancy may well be more important than genetic influences. Growing tall and strong in early life and young adulthood has many competitive advantages in pursuits and occupations where strength is required, and hence in general this must be a beneficial feature for farmers.

If this early height and often-associated weight gain is due to food intake, then there may also be harmful effects which will only be noted in later life. In a study conducted in Northern Ireland we found that the average fifty-year-old farmer was 5 ft 8 in and weighed 12 st 5 lb. This is at least 14 lb heavier than the ideal weight for this height. It is certainly not uncommon to see older farmers with considerable obesity, especially when family members start to take over some of the physical tasks of the farm. Thus the early advantages of growth in the young farmer may well become disadvantages in later years if a calorie intake out of proportion to the calorie expenditure is maintained.

Obese people tend to die younger than those of average weight. With obesity goes the risk of arterial disease and hence angina, heart attacks, high blood pressure and strokes. Blocked arteries to the legs may also develop giving leg pain and possibly resulting in gangrene. The incidence of the late onset and weight-related variety of diabetes is increased and this will also contribute to arterial disease. Joints are put under unnecessary strain and the normal process of ageing of the cartilages is accelerated with the development of osteoarthritis, especially of the knees.

	OBESITY
Check:	**What your weight should be for your height.**
Side effects:	**Diabetes, heart attacks, stroke, high blood pressure, cramps in legs, gangrene, arthritis.**
Prevention:	**Diet. Reduce cholesterol and fat intake. Exercise regularly.**

Farming in worldwide terms is still a very labour intensive occupation. However, like other industries in the more developed countries, the amount of direct human physical activity has been greatly reduced by labour-saving devices.

In cities people frequently take buses or drive cars rather than walk. Cooks use electrical food processors for baking rather than beating mixtures with wooden spoons. On the farm, ploughing can be done while sitting listening to a radio in a heated tractor cab. Milking is just a matter of transferring the cluster from one set of teats to the next. Using the tractor with a forklift is certainly a much less active process for moving hay or silage than using a pitchfork. Thus although physical activity is still required on the farm, it is generally much less intense and for shorter periods of time than previously. Despite this increase in energy expenditure, food intake has been maintained or increased in parallel with other changes in living standards in the Western world. Thus there is the risk that energy intake will outstrip energy usage. It therefore can no longer be assumed that a farmer should be healthier than city dwellers because of the nature of his work.

Industrial workers and professional people often get their exercise by participation in outdoor sports or, increasingly, at sports or leisure centres. Farming, although perhaps not as energetic a job as before, is still very time consuming and requires the farmer's presence every day. There are thus few weekends or days off and this pattern does not easily lend itself to regular participation in organised sport or non-work-related exercise. There must, however, be few farms where better organisation, cooperation and a determination to participate would not allow time off to be made available. This is very important not only for the farmer himself but also as an example for his children. If time away from the farm cannot be arranged, it is always possible to walk at a faster pace in the fields or lanes. Likewise some of the more physical tasks could be done by hand in a competitive way even if mechanical methods are available.

Increase leisure exercise.

The advantages that come with a healthier diet and physical exercise may be lost if the fact that cigarette smoking is a serious health hazard is ignored. In our recent study we found that 41 per cent of Northern Ireland farmers had never smoked compared with 34 per cent of other similarly aged rural dwellers. This ratio is similar to that of Vermont, USA, and thus the farming community is setting a good example. We also found that 29 per cent of the Northern Ireland farmers were still smoking cigarettes, although many former smokers had stopped. People who smoke are much more susceptible to the development of chronic bronchitis, emphysema, lung cancer and all arterial diseases. Many governments and local authorities are now introducing non-smoking policies in public areas of buildings and transport systems. Farmers, by the nature of their occupation, are less likely to come under the influence of these measures and are thus in danger of slipping from their place in the non-smoking tables.

SMOKING

Smoking causes cancer, bronchitis, emphysema and arterial disease.

It is never too late to stop.

The other factor that applies to farmers just as much as the rest of the community is the consumption of saturated animal fats in food. These fats are a risk to some people beyond the risk of simple obesity. It is natural that producers of meat, milk and eggs consider these foods to be wholesome and good. However, they all contain a fatty substance, cholesterol, which is a constituent of all animals and is necessary for their existence. When not acquired from food, it may also be produced within the body from other food substances. The level of cholesterol and other fats in the blood depends upon a balance between food intake, the body's own production, the amount stored in the tissues and the rate of metabolic breakdown. Although intake is important, some of the other factors are probably genetically determined and are therefore not open to dietary manipulation. In all people, however, the level of cholesterol can be reduced if the dietary intake is low.

The reason for concern about cholesterol and other animal fats is that high levels of it in the blood are associated with arterial diseases. Thus farmers with a family history of heart attacks, strokes and arterial disease of the legs, especially at an early age, may have genetically high levels of blood fats. If there is such a history, then it would be worth getting the blood cholesterol level estimated. If this is high, specific advice is available. Vegetables and vegetable products are low in cholesterol and it would seem sensible for those living in the more developed nations to substitute some of these in place of animal products. Recently I discovered after a lecture given to farmers in Northern Ireland that none of them had tasted skimmed or semi-skimmed milk, used vegetable oil for cooking or tried low-fat spreads for bread. Coincidentally, few of them had eaten high-fibre breads. Thus data about the relationship between diet and health seems to have bypassed them, unlike their urban neighbours. Farmers in less-developed countries eat diets containing less animal fat and higher fibre content out of both choice and necessity.

It is of interest that people with very low cholesterol levels have an increased mortality rate from cancer. Perhaps since some animal products are rich sources of proteins, vitamins and minerals, we should all take a commonsense approach and modify our diets to get the benefits from both animal and vegetable foods eaten in moderation.

CHOLESTEROL

This is an essential constituent of the body.

High levels in the blood are associated with disease.

Family history of heart attacks in the young is a risk factor.

Balance animal with vegetable food intake.

It has been known for some years now that diseases of the large bowel, including diverticulitis and cancer, are more common in the Western world than in Africa. There are many possible explanations for this but it is probable that diet plays a major role. It is known that the unabsorbable or undigested food residue (faecal bulk) is much greater in the residents of those countries where such diseases are low. The best way to increase this bulk is to eat vegetable foods with a high fibre content, e.g. bran. This can be done either by eating relatively unprocessed foods, such as brown bread and brown rice, or by adding bran to processed foods. All these, like meat, are of course grown by the farmer for the community; why should he not benefit from eating them?

Arthritis and low back pain are amongst the most frequent

7

Saturated fat and cholesterol content of some common foods

Food	Saturated fat g/100g	Cholesterol mg/100g
MEAT AND FISH		
beef stewed or minced	6.4	82
grilled rump	5.1	82
roast sirloin	8.9	82
stewed liver	3.5	240
tongue	7.8	270
sausage	7.0	42
burger	7.3	68
lamb roast leg	8.7	110
grilled chop	14.1	110
breast	18.1	110
pork grilled chop	9.6	110
roast leg	7.8	110
fried streaky bacon	18.0	110
boiled gammon joint	8.0	110
boiled chicken	2.4	90
white meat	1.6	80
roast turkey	2.2	80
white meat	0.5	62
fish cod	0.2	60
plaice	0.3	90
herring	2.6	80
mackerel	3.9	80
DAIRY PRODUCTS AND COOKING OILS		
cream single	12.6	66
double	28.8	140
milk whole	2.3	14
skimmed	0.04	2
cheese cheddar	20.0	70
cottage	2.4	13
butter	49.0	230
margarine hard	30.0	variable
vegetable	25.6	trace
polyunsaturated	19.0	trace
vegetable oil maize	16.4	trace
olive	14.0	trace
lard	41.8	70
peanut butter	10.6	trace

Food	Saturated fat g/100g	Cholesterol mg/100g
CAKES AND PUDDINGS		
fancy iced cakes	9.28	120–260*
rich fruit cake	3.9	40
cheese cake	18.7	95
sponge cake	2.1	260
sponge pudding	5.9	80
dairy ice cream	4.3	21
shortbread	15.0	120–260*
chocolate digestive	12.0	120–260*
meringues	0.0	0
milk chocolate	60.0	47
plain chocolate	60.0	trace
OTHER FOODS		
boiled egg	3.4	450
quiche lorraine	12.5	130
hot pot	1.8	25
bread	0.4	trace
avocado	2.6	trace
banana	0.1	trace
potato boiled	0.01	trace
rice	0.0	trace

* Variation in quantity due to different recipes.

complaints that medical doctors are presented with. They are very common in industrial settings and frequently cause work loss. The exact cause of most joint diseases is not known but they can be divided up into four main groups: osteoarthritis, infective arthritis, rheumatoid arthritis and metabolic arthritis (gout, etc.)

Osteoarthritis tends to be due to wear and tear and it is common in older people in general. It especially affects the hips and knees. Other joints tend to suffer if they are damaged or subjected to repetitive stress. People who use their hands in manual work tend to get arthritis in the finger joints with resultant gnarled hands from swollen and distorted knuckles. This is not uncommon in farmers. Arthritis of the spinal column is common in those who have to lift heavy weights. This is so common amongst farmers that in some regions a local name such as 'Norfolk Back' might be used. The increased mechanisation on farms should gradually reduce the incidence of this condition in farmers. It is essential, however, that farmers use the best and the recommended methods for manual lifting. To do this the back should be kept straight and as vertical as possible so that most of the bending takes place at the hips and knees. Stress on the spinal column may also damage the soft discs between the vertebral bones. Rupture of the outer support of the disc may cause the inner pulp to protrude and compress the surrounding nerves. If the sciatic nerve is damaged, severe pain develops in the lower back and down the back of the leg.

BACK STRAIN

Lift weights properly.

Prevent arthritis of the spine.

Prevent vertebral disc rupture.

Infective arthritis results when germs enter the joint from the bloodstream or directly through a penetrating wound. The farmer is exposed to many infectious agents which might cause arthritis. Special types of infection related to work with animals are dealt with in the chapter on zoonotic diseases (Chapter 8). Other common varieties of infective arthritis caused by blood-borne germs are probably no more common in farmers than the rest of the community. Farmers, however, are more susceptible to joint injury which may give access to bacteria. All forms of infective arthritis are particularly acute and require urgent therapy.

Incapacity, of course, has financial consequences for the self-employed farmer. Early medical attention is thus important if the cause of the arthritis is to be diagnosed and treatment initiated before further secondary damage takes place. Any arthritis is made worse in the acute stage by the physical nature of farm work.

ARTHRITIS

Joint pains may have a treatable cause.

Consult your doctor early.

I have already mentioned hobbies in terms of physical exercise. They are also important because of the psychological break they give from the day-to-day routine of farm work. Farming is frequently a 24-hour-a-day job, especially during the lambing and calving seasons, when added night-time work increases the mental as well as the physical stress on both the farmer and his family. Radio and television have increased the contact between farmers in relatively isolated regions of developed countries and the rest of their community. However, these forms of entertainment are essentially passive and may even reduce communication skills by inhibiting conversation. The ability and opportunity to get involved in non-sporting hobbies and even reading may be reduced because of educational reasons and lack of interpersonal skills. There is a tendency, in some regions, for the children of farming families to miss school during the busy periods and to leave their formal education at the earliest possible time. This prevents the children from attaining their maximum educational potential and thus they will be at a disadvantage compared with their peers in towns.

Depression and suicide are as common in farmers as in the rest of the community. This may be partly brought about by the relative isolation and the factors mentioned above. Access to caring organisations such as the Samaritans in the UK may be limited by lack of telephones, lack of knowledge of their existence or by inhibition over talking about personal affairs to people from a different background. Besides, farmers are renowned for keeping information 'close to the chest'. Fortunately, market attendance and religious observance often give the opportunity to communicate personal problems even if only indirectly, and attendance at these events should be encouraged. Heavy alcohol consumption is not a realistic alternative option to these measures.

Alcoholism is no more a problem among farmers than their rural peers. Intake may, however, be generally high in those regions where wine, cider and spirits are produced. There is the added burden to the family if a farmer is not able to cope with work or even losses to the farm as a result of drinking to excess.

MENTAL STRESS

Be careful not to become isolated.

Have at least one hobby.

Do not turn to alcohol for stress relief.

The above are examples of types of illness that farmers and their families may suffer from in the same way as the rest of the community. Farmers are probably less likely to be hypochondriacs than their peers and it is not the intention of this book to make the reader introspective. There is, however, a tendency to accept some illnesses as part of the job and hence not to do anything about them. Farmers should realise that there is a difference between complaining and seeking a remedy for an illness. It is usually better to take advice earlier rather than wait until irreversible harm has taken place.

2

Farming Accidents

The Health and Safety Statistics for the United Kingdom in 1986 showed that the average incidence of industrial death for all occupations was 65 per 100,000 workers. This was an increase of 40 per cent over the previous four years. For agricultural workers the incidence for 1986 was 77 per 100,000 workers, which was also an increase of 40 per cent over the same period. The statistics also showed a 50 per cent increase in major non-fatal accidents related to farm work. Thus, from the point of view of accidents, the farm is a dangerous place to work.

> **There are more deaths and serious accidents on farms than in the average industry.**
>
> **The situation is getting worse.**

Farmers and their families are just as likely to have general accidents as anyone in the rest of the community. A farmer's child is as likely as a city dweller's child to pull a pot of boiling water over himself or herself. Someone can fall downstairs in the farmhouse just as easily as in a city estate block or house. Thus vigilance to avoid this type of possible accident is just as necessary on the farm.

Car accidents are frequent in all communities. It is especially tragic, however, when a young farmer accustomed only to the security of the slow speed of his tractor or, more likely, irritated by the slowness of farm vehicles, kills himself in a fast car as soon as he gets a few hours' break from the farm. Most accidents to farmers and their families take place around the farmyard, but since they also happen in the fields, rivers and roadways, vigilance is always required.

13

This chapter outlines some of the problem areas for accidents. Each individual farmer must, however, be alert to the safety problems on his own farm in the same way that a factory manager or worker has to be responsible for his or her sector in industry. Government regulations vary from country to country and the reader must be familiar with them.

ACCIDENTS

The farmyard is an especially dangerous place.

You are responsible for your own safety.

Learn the government regulations.

Tractors

Accidents with tractors and tractor-powered equipment are very common, yet most of them could be prevented. Fewer accidents occur nowadays from driving horizontally across hill fields, but they may happen if care is not taken when turning at the top of a hill. Likewise, vehicles may roll over if turned too quickly on flat ground. With foresight these accidents could mostly be avoided.

It is false economy not to have some sort of protection for the tractor driver and at the very minimum a roll bar should be in place. In the UK it is mandatory for a safety cab to be on the tractor before anyone under the age of 16 is allowed to drive. It is also illegal for a child under the age of 13 to ride on any part of the tractor, even if a cab is fitted. Several farmers have been prosecuted for breaking this law.

TRACTOR ACCIDENTS

Drive all vehicles with care.

Fit safety roll bars and cabs.

Do not allow young children on tractors.

An unloaded tractor can go up and down inclines as steep as 25–30 degrees as long as the ground is firm and dry. Loaded with a 1000-gallon slurry tank, however, a two-wheel-drive tractor may in some conditions slip on a slope of only 6 degrees and a four-wheel-drive on a slope of 9 degrees. A tractor can take equipment with brakes up a hill as steep as 21 degrees.

THE TRACTOR ON HILLS

**Check breaking safety limits on slopes
for all farm vehicles.**

Tractors can tip forward or backward if too heavy a load is lifted. Once the tilting begins it can take less than one second for the tractor to lift completely off the ground. Since this gives little time to think, it is important to have a clear plan of action for such an emergency. In general it is better to grip the wheel tightly and depend on the safety cab rather than try to jump clear.

THE TRACTOR AS A FORKLIFT

Do not exceed the recommended lifting weights.

Plan your reaction to tilting accidents.

If passengers have to be transported on a tractor, then a safety cab must be fitted and the passenger should be carried in as safe a position as possible. They must never be carried on the linkage or drawbar. No one should be carried on towed equipment (or digger shovels) unless it is specifically made for passengers. Children (or in fact adults) should not be carried on loaded or low-sided trailers. One slip may, and in fact frequently does, cause death.

15

PASSENGERS ON TRACTORS

Never carry children under the age of 13.

Never carry anyone unless in a safety cab.

Loaded trailers are death traps for passengers.

General advice about motor vehicle safety applies as much to the driving and parking of tractors as of cars or buses. The Highway Code must be adhered to. Special care is needed when taking tractors from fields onto public roads even when signposts have been placed warning other road users.

An especially dangerous time in the UK is when taking a right turn into a field. Drivers behind a slow-moving farm vehicle may well get frustrated and take any opportunity to pass. If the tractor driver does not make a recognised and clear signal that he is going to turn, then an accident may ensue. Even if the tractor driver does indicate, some road users may still overtake. The tractor driver must never assume, therefore, that cars coming up from behind will act on the signal, and must be prepared to take avoiding action if necessary.

All these accidents are more likely to happen if the tractor driver is distracted by the use of headphones. Ear protection is needed with some vehicles, and if worn, the driver must become more alert to visual signals of impending danger.

As with other farm machinery, the moving parts of the tractor, such as the power take-off shaft, must be covered along the whole of their length. Direct contact is obviously dangerous, and if an article of clothing blown by the wind gets caught up in the machine, it can pull the wearer to his or her death.

Bystanders are also at risk when mowers, harvesters, hedge cutters and manure and fertiliser spreaders are in use. They need to be kept well out of the way of potential missiles such as stones or branches. This may become a greater problem in areas where the population is increasing or on farms which incorporate tourist or educational attractions. Linking and unlinking equipment is another time of high risk and help may be needed if crush injury is to be avoided. Supports must always be stable and not easily knocked over. Apparatus for attachment to tractors is becoming increasingly refined and no specialist equipment should

16

be used without first seeing a demonstration or at least reading the literature carefully.

Cover all parts on a tractor.

Keep bystanders clear of missiles.

Linking equipment may be dangerous without help.

Hand-held chain saws are frequently used for hedge cutting and forestry work, and they present the obvious risks of cutting the limbs of the operator or his helper. A less obvious risk is that of Raynaud's phenomenon, otherwise known as vibration white finger or episodic blanching. Vibration causes the arteries to one or more of the fingers to contract, cutting off the blood supply to them. The finger becomes painful and subsequently as the blood supply returns, it turns blue and red with the development of an intense 'pins and needles' sensation. The problem is more likely to occur in cold weather.

Other risks when using the chain saw include eye damage from flying splinters, ear damage from the noise, skin burns from touching the hot areas and injuries from falling trees. The user must always work from a position of stability and not be liable to overbalance. Proper protection should also be used, especially for the eyes, ears, legs, hands and feet.

Fixed Equipment

In the context of farming, fixed equipment includes not only mechanical and electrical apparatus but also the building, farmyard and walkways. They must all be properly maintained and equipment guarded to the same standards that would be expected in any industry.

All the moving parts of machinery need to be guarded to prevent accidental contact with passers-by.

Keep equipment well maintained.

Keep the farmyard tidy.

17

The ground in working areas must be well lit and free from objects such as brushes, buckets, sacks and stray animals. Any of these may cause the farmer to fall or a tractor to swerve suddenly and topple over. All stairs should be fitted with a hand rail.

Most floors are adequate for the purpose they were originally intended for; however, as the farm modernises, areas may be put to new uses. Floors should be assessed and strengthened if necessary, especially in the region of slurry pits before additional weight is applied.

Silage

Silage clamps must be surrounded by strong walls and end barriers if tractor and other accidents are to be avoided. Professional advice should be taken before building or reinforcing a silo.

Tower silos pose special risks. Firstly there is a risk of falls from access ladders and secondly there is the problem of poisonous gas formation. For both these reasons, access to the silo must be restricted to those who understand the problems and who have read the literature. This aspect is dealt with in more detail in Chapter 5. It is important to remember that because of the heavy toxic gases, adequate ventilation is required **before** entry to the silo, and several strong colleagues should be available to hold a safety harness if necessary.

SILAGE

Check walls and ends of clamp silage pits.

Tower silage stores may kill by falls or poisoning.

Slurry

Slurry tanks are an area of high risk of accidents such as drowning, poisoning, fire and explosion.

All slurry lagoons and tanks should be surrounded by a 1.8 metre high fence with childproof gates, and children must be warned not to enter these areas. Manholes must be equipped with a heavy cover and also fitted with a secondary guard to prevent accidental falling during access to the tank. When slats are removed, temporary covers or rails should be installed to protect the opening.

SLURRY TANKS AND LAGOONS

Prevent drowning: Check manhole covers.

Erect fences and lock gates.

Place guards around open slats.

Details about gases are given in Chapter 5. Gas release can be reduced if surface agitation is kept to a minimum, especially when silage effluent is added. The concentration of the released gas is further reduced by good ventilation, which, although relevant to any situation involving poisonous gases, is especially important before entering an empty tank. Even then, the farmer should only enter if there is help from two other strong adults holding the ends of a light harness from outside the tank.

Smoking and naked flames should be avoided in the vicinity of a slurry tank in view of the risk of fire or explosion.

SLURRY TANKS AND LAGOONS

Avoid poisoning: Reduce surface agitation.

Ventilate the tank before entry.

Use a safety harness.

Avoid explosions: No naked flames near slurry.

Ladders

Ladders must be strongly made and checked for faults each time before use. Ideally the base of the ladder should be placed 1 metre out from the wall for each 4 metres up the wall, and the top should extend 1.1 metres beyond the place for stepping off. If there is an extension to the ladder, the correct overlap must be used, i.e. a two-rung overlap for up to 5 metres extension, a three-rung overlap for 7 metres extension and a four-rung overlap for more than 7 metres.

All ladders must be held, strapped or otherwise safely secured at the base before climbing. If you have to stretch out to the side while on the ladder, it should be secured at the top as well. Extra security is required in high winds or if animals or moving vehicles are near the base of the ladder.

Check ladders for faults before use.

Secure top and bottom.

Use correct overlap for extension.

Electricity

All electrical equipment is potentially dangerous and should have a switch-off point near the area of use in the event of an emergency arising.

All electrical circuits need to be properly installed and the equipment earthed. Malfunctioning circuits and equipment should be professionally checked. Waterproof plugs and switches are an advantage on a farm and an earth circuit leakage breaker (earth trip) will give additional safety.

All extension leads are potentially dangerous and especially on a farm, where they may fall into water or get damaged by animals or tractors.

Electrical circuits should be checked.

All equipment must be earthed.

Use earth circuit leakage breakers.

Do not use extension leads.

Chemicals

All chemicals, no matter how common or frequently they are used, may in certain circumstances be poisonous. If this general principle were remembered, there would be fewer cases of poisoning. Only those chemicals which are approved by the Ministry of Agriculture, Fisheries and Food should be used on farms. A chemical's toxicity depends on the formulation and strength of the manufacturer's individual brand. Since this can change from time to time, always check the instructions even if you are familiar with the basic ingredients.

The degree of protective clothing required when working with chemicals depends on the chemical used and its formulation. Rubber gloves, a face shield, protective clothing and boots should be used as indicated. A supply of clean water should always be available in the vicinity of chemical use to allow for washing in the event of accidental spillage. An emergency plan should also be prepared to avoid confusion at the time of an accident. The necessity of washing hands and other exposed parts before eating (or smoking) cannot be emphasised enough.

All chemicals may injure or kill.

Always read the manufacturer's literature.

Use adequate protection.

Wash thoroughly after use.

All chemicals, regardless of how innocent they may appear, should be locked in a childproof (and adult-proof) store.

Ideally chemicals should be kept in the original container with its labelling: if they do have to be transferred (e.g. due to the original package being damaged or developing a leak), they should **never** be stored in an unlabelled container.

CHEMICAL STORAGE

Keep chemicals locked up.

Keep them in their original container.

Animals

Animals, like humans, are all individuals, and even those who by temperament are generally placid may at times become aggressive. It is because of this lack of predictability that all animals should be considered potentially dangerous. The degree of risk partly depends upon the relative size of the animal compared to the human, since a small animal may not be a danger to an experienced adult but might injure a child.

All bulls should be ringed by the age of ten months and a stick should always be carried when walking amongst cows and heifers. Boars and sows must be managed with caution and a handling board used to avoid leg injuries.

ANIMAL HANDLING

Some animals are *always* aggressive.

Most animals may *occasionally* be aggressive.

Use a crush, handling board or stick.

Guns

Many farmers have shotguns to control birds, vermin and other wild animals. Guns must always be kept in safe custody and away from children. They should never be carried ready loaded, and they are especially dangerous when travelling on machinery or when climbing.

Keep guns clean and locked away safely.

Never carry a loaded gun on machinery.

Never climb with a loaded gun.

General Safety

Different parts of the body require protection when special tasks are performed on the farm. Too often these difficult tasks become commonplace and their safety problems are forgotten. Familiarity breeds contempt for the dangers, and bravado and the need to get things done quickly tempt the farmer not to use protective wear. This is false economy since it may lead to loss of life or injury sufficient to cause work loss.

Eye shields must be worn if there is a risk of toxic or caustic liquids being splashed. Likewise, dust particles and thorns may get into the eyes if they are unprotected. In a factory the workforce would be dismissed for failing to comply with the rules concerning eye protection. Blindness from natural causes is bad enough without the additional risk from the work place.

EYE PROTECTION

Use eye shields.

Protect from splashes and solid particles.

The farmer is frequently exposed to a dangerous level of noise (defined as above ninety decibels). Chain saws and tractors can be very noisy (like some animals!). Ear protectors should be used, especially during prolonged exposure to these noises.

It should be remembered that headphones playing loud music may be dangerous, not only because of the additional noise but also because they can distract the wearer from the task being done. Safety cabs of tractors will also reduce noise provided the doors are kept shut. The degree of protection which cabs and ear protectors offer should be checked before purchase.

As farmers grow older they develop diseases of the inner ear in the same way as other members of the community. These diseases often result from virus infections or arterial disease. The end result is usually decreased hearing and a feeling of light-headedness or a buzzing noise in the ears. Age and noise damage can cause considerable deafness that is dangerous in an industrial setting. Farmers suffering from light-headedness or dizziness should not climb heights or drive.

EAR PROTECTION

Many farm animals and equipment are noisy.

Protect against deafness.

Trauma

Only after an injury is a person aware of most parts of the body. This classically applies to the fingers and toes, injury to which leads to pain and inconvenience beyond their size and they must be protected. Even minor abrasions may injure the skin sufficiently to allow the entry of germs. Thus conditioning (barrier) creams and gloves should be used when indicated (see Chapter 3).

All manual workers are susceptible to physical trauma and farmers are no exception. Causes can range from minor cuts and bruises to broken bones, major internal bleeding, concussion and death. If severe damage is suspected, the farmer should be kept still and warm while waiting for help. Since surgery may be needed, food and fluids (including alcohol) should not be given. Bleeding can usually be stopped by applying pressure directly over the bleeding point with or without the help of a clean cloth. If a limb has been severed, it should be kept cool and sent to hospital with the patient.

If you have not had first aid training, find out from your doctor or local government agency where a course is available. Such training may save a life.

Tetanus results from germs entering abrasions and cuts and is described in a later chapter. It is, however, important to emphasise in this section on trauma that all farmers and their families should be immunised against tetanus and have their level of immunity boosted every few years.

Accidents can often be prevented by considering the possible risks in advance. Farmers often jump down from tractors or trailers and strain their ankles by overextending the joints. This is more likely if there are loose objects or stones on the surface. Care must be taken to avoid this form of accident since it may lead to considerable disability. Safety boots should be worn if you work regularly with objects that are likely to fall on the feet.

Prevention is better than cure.

Step down from heights – don't jump.

Learn first aid.

Immunise against tetanus.

Children

Twenty-five per cent of all farming accidents in Northern Ireland involve children under the age of 16. This proportion is not uncommon in other farming communities. By far the most common cause of death is being caught or crushed by machinery. Many children also die from drowning in slurry pits, in farm fires or from contact with poisonous chemicals.

Farms are and always will be homes but they are also industrial sites. Factory owners would be prosecuted for allowing children into dangerous areas on urban industrial sites and the same general principle should apply to farms.

CHILDREN

Do not use them for cheap labour.

Do not allow them on tractors or machines.

Give them a safe place where they can play.

Lock up all poisons and guns.

25

I recently followed behind a digger which was being driven in a narrow country road with three children and a bicycle in the back shovel. They were having a great time but any or all of them could have been killed during their journey. Advanced driving instructors teach that good driving is an attitude of mind on the part of the driver, a matter of taking pride in driving properly. It is only when farmers take such an approach to safety on their farms that the death and injury rate will decline from its current unacceptably high level.

3

Climatic Risks

Farmers, like other outdoor workers, are at increased risk from the extremes of climate. A few of these risks are dealt with here, and in most cases the use of simple commonsense measures should prevent any problems.

Rain and Water

Rain is needed for the growth of crops and thus in moderation is welcomed by farmers. An occasional 'wetting' when out in the fields is no more harmful than going for a swim or shower. There may, however, be problems under certain circumstances.

Most farmers know their local rivers and the areas that are safe to wade in. When excess rain cannot be drained, flooding takes place and the nature of the underlying terrain is not so obvious. It is at times like these that many farm animals may drown and the farmer may also suffer a similar fate while trying to rescue them. This may also happen if he tries to do emergency drain clearing under less than perfect circumstances. Flood waters thus should not be entered without safety precautions.

FLOOD WATER

Clean drains in dry season.

Take care in flood water.

Do not allow skin to become soaked.

27

Working in rivers, lakes and flood water brings the possibility of coming into contact with a number of infectious agents. These are dealt with in Chapter 7. The infection depends upon the country, but in most countries rat-borne diseases are the most common type.

The skin is a good natural barrier against penetration by infectious agents; however, skin that has become soft from prolonged contact with water no longer provides such a barrier. Also, if there are scratches or lacerations on the skin, then the risk from disease following contact with any germs or toxic chemicals is greatly increased.

If you know your hands will be immersed for a prolonged period, use a good conditioning cream. Gloves are helpful only if the insides remain dry.

The presence of puddles of water increases the risk of slipping for humans, animals and machinery. Extra care is therefore necessary in wet weather. In warm countries, pools of water are ideal breeding grounds for mosquitos. This increases both the risk of skin irritation from bites and the transmission of malaria and other infections.

Cold

Most deaths at sea are due not to drowning but to hypothermia, which is an abnormally low body temperature. Hypothermia can also occur, even in the young, following prolonged wetting of clothes by rain. Heat loss is increased by the evaporation of the moisture and the loss of the clothes' insulating qualities, a problem which is accentuated in windy weather. It is thus important to change out of wet clothing as soon as possible.

Although the consumption of alcohol makes the skin feel warmer, heat loss from the body is in fact increased. It is this central or core cooling that leads to a deterioration of mental alertness, a slowing of the heart and eventually a loss of consciousness. This can be very serious in the older farmer, who might already have heart disease or blocked arteries to the brain. Severe cooling, even in the young, may also lead to the development of pneumonia.

HYPOTHERMIA

Avoid hypothermia by keeping dry.

Protect from cooling winds.

Cooling not only leads to a loss of central body functions but may also cause local disease. Frost bite is an obvious example and may take place rapidly, especially if the circulation is already slowed by other disease. The arteries of some people are exquisitely sensitive to cold and respond by going into spasm. This reduces the blood supply to the skin of the fingers, which may become white and painful and subsequently blue. When the hands are warmed, the fingers become red with a painful tingle. This phenomenon is known as Raynaud's disease. The best way of avoiding it is by the use of good insulating or warmed gloves.

Sun and Heat

While there are obvious pleasures in sun and heat, there are also a number of potential dangers. The consequences of these will vary from country to country.

Sunlight normally leads to tanning of the skin due to the effects of ultraviolet light. An excess of this ultraviolet light (a risk which may increase due to the effect of chlorofluorocarbons on the ozone layer) can produce cancer of the skin. This is more common in fair-skinned people and tends to develop on the face, head, backs of hands and forearms, since these areas are most exposed. The cancer may look like only a patch of dry, hard-crusted skin. If the top falls off, as it does from time to time, it will recur and may look like a small sore. Such cancers, if treated early, are completely curable and should be brought to the attention of a doctor when first noticed. They are, however, best avoided by reducing exposure to a minimum by using clothing, hats and barrier creams or by erecting sun shades. A melanoma is a more severe form of skin cancer that often develops from a mole, and these are also more common on sun-exposed areas of the body.

SKIN CANCER

Reduce direct sun exposure.

Seek medical advice early.

29

The body produces sweat, which evaporates and cools the surface of the skin when a person is working hard or subjected to high atmospheric temperatures. Sweat is salty and thus excess loss of both water and sodium in the sweat may occur, especially on hot days. If sufficiently prolonged this may lead to the development of heat stroke with weakness, a drop in blood pressure and altered consciousness. The effect on blood pressure may be even more marked if the farmer is taking drugs that affect the fluid output from the body or drugs which can by themselves lower blood pressure.

Heat stroke is best avoided by ensuring adequate salt and water intake in very warm weather. Add a little more salt than usual to your food (unless your doctor has instructed otherwise because of high blood pressure or heart failure). Drink to quench your thirst and check that the colour of your urine does not become very dark due to lack of water.

During the spring and summer large amounts of pollen are produced by plants. Pollens, especially those from grass, are notorious for causing allergic reactions in some individuals. An allergic reaction develops because the pollen substance (antigen) incites the blood cells of the body to form antibodies which circulate in the blood and become attached to body cells. When pollen or other antigenic substances subsequently come into contact with the special antibodies, chemicals are released which cause the observed symptoms. The reaction may be one of 'hay fever' with the symptoms of itchy, watery eyes and a running nose with sneezing and blockage. There may also be skin blotches which look similar to nettle rash. A cough with sputum and wheezy shortness of breath (asthma) may develop. Despite its name, hay fever is rarely associated with a raised temperature.

It is obviously difficult for a farmer to avoid pollens because of his outdoor occupation, although the cutting of early grass for silage may reduce some of the local concentration. There are now a number of drugs and other methods available which can greatly reduce the symptoms. It is therefore worthwhile consulting your doctor before the symptoms become so severe that work is affected.

Allergic symptoms can be reduced by simple measures.

Some of the above mentioned symptoms may develop at any time and can last all year (that is, they are perennial). This is because of the presence of non-seasonal allergens and other factors. Similar forms of therapy are also available for these conditions.

During the autumn, winter and early spring the farmer is more frequently indoors. This brings him into closer contact with some of the farm animals and hence an increased risk from accidents, infections and allergic reactions to foodstuffs. These matters are dealt with in subsequent chapters.

4

General Health

This chapter is not meant to be an encyclopedia of medical conditions. It is intended to give a few examples of some of the common ailments which might afflict any member of the community. If they attack a farmer, his work will suffer. The most important or prevalent diseases in any community will vary from country to country. Thus in poor and relatively underdeveloped countries, the main problems may be related to malnutrition and infection due to poor sanitation and contaminated water supplies. In the more affluent parts of Europe, North America and Australasia, diseases related to excess food intake, self-induced injury and cancer are the most important.

Farmers often accept symptoms as part of their job and do not tend to seek advice until the disease has progressed to a more advanced state than they would allow in their animals. The few examples of diseases given here may help the farmer to recognise the common diseases earlier, understand the consequences and seek help if necessary.

Arterial Diseases

A narrowing of any artery reduces the amount of blood that can flow to a tissue or organ. If the narrowing is severe, symptoms will be produced in the tissue or organ when it is at rest. If it is not so critically narrow, then symptoms will only occur during exercise when the demands are increased beyond what can be supplied by a narrowed artery. When narrowing is present, there may be clotting, which can completely stop the flow of blood. Symptoms produced will depend upon the organ supplied. Complete stoppage will obviously result in the death of the organ, which, if critical, may result in the death of the person.

Blood is supplied to the heart by the coronary arteries. Many people in Western societies have some disease of their coronary arteries by the time they are twenty years old. In some families, the symptoms are present even at this age, but it is more usual for them to begin after the age of forty or fifty. The heart muscle, as with other muscles, develops pain when insufficient oxygen is given to it. Since exercise causes the heart to beat faster and to need more oxygen, symptoms develop with effort and the pain is called **angina**. It may be a pain or a sensation of tightness or crushing across the front of the chest or occasionally in the neck or arms. It tends to disappear after five to fifteen minutes rest. If such symptoms develop, then a doctor should be consulted.

If the narrowing continues, heart muscle fibres, supplied with oxygen by the artery, die and are replaced by scar tissue. This area will not pump as efficiently as the undamaged muscular area, and blood will accumulate in the lungs and body, leading to shortness of breath and swelling of the feet and ankles. These symptoms, like angina, can be treated. A **coronary thrombosis** or clot leads to a **heart attack**, which usually results in the sudden death of a segment of heart muscle. This presents itself as a more severe central chest pain which does not ease with rest and is often accompanied by weakness, sweating and nausea. If this occurs, it is better to stay still and get help to come to you. Never continue to drive or work. Quick treatment limits the degree of damage to the muscle and should allow you to return to work sooner.

If the arteries to the legs are narrow, then the leg muscles, especially those of the calf, become crampy and sore when walking. This is called **intermittent claudication** and may require surgical correction to open up the arteries.

The arteries to the brain may also become narrowed and if this is a gradual process, it may result in a gradual reduction in mental function. Occasionally the sufferer notices excess light-headedness or even a loss of consciousness when the head is turned round to look over the shoulder. This movement results in kinking of the blood vessels and further loss of blood flow. If the blood vessels are narrow, a clot may develop within them resulting in damage to a section of the brain. This is known as a **stroke**, and it usually results in loss of power in an arm, leg or one side of the face. Difficulties in speech and sight may be noticed and, if more severe, consciousness may be lost. Narrowing of one of the large arteries in the neck may lead to dizziness or even loss of consciousness when the head is turned, e.g. when reversing a tractor.

If arterial disease is diagnosed at an early stage it may be possible to reverse some of the changes or slow their progression.

High blood pressure (hypertension) damages the arteries in the body and thus makes the above diseases possible at a younger age. Because of the raised pressure in the arteries, weak areas may rupture with resultant bleeding, e.g. a brain haemorrhage. This high pressure may also damage the kidneys, while stress on the muscles of the heart leads them initially to enlarge and ultimately to fail. Measurement of blood pressure is a simple, quick and accurate test and all adults, including farmers, should have this done occasionally. Treatment for hypertension is now excellent and will prevent the development of serious complications.

With all these arterial diseases, the main thrust must be towards prevention rather than cure (see Chapter 1).

PREVENTION OF ARTERIAL DISEASE

Do not smoke.

Keep weight normal for height.

Reduce cholesterol intake.

Take exercise.

Have blood pressure checked.

Respiratory Disease

The main functions of the respiratory system are to clean the air we breathe, bring it in and out of the lungs through the windpipes (trachea and bronchi) and allow air to come into contact with the blood through thin membranes in the spongy part of the lung (alveoli), where the gases are exchanged.

Everyone produces water in the nose and bronchial tree to moisten the breathed-in air and prevent unnatural drying. Mucus is also produced to trap particles of dust and germs. This mucus is normally carried by minute hairs (cilia) to the back of the throat, where it is swallowed with the saliva from the mouth. Irritation of the bronchial tree from foreign substances or disease leads to an increased production of these fluids, which are then coughed up as sputum (spit or phlegm).

Chronic bronchitis is characterised by a persistent cough for at least three months of the year over several years. Most population studies show that this condition is present in 6–18 per cent of the adult population. In a rural community in Northern Ireland, 18 per cent of the population had bronchitis, but this rate was only 6 per cent in those who had never smoked. There is little doubt about the relationship between smoking and this disease, and symptoms tend to disappear in those who have ceased the habit.

CHRONIC BRONCHITIS

Prevent chronic bronchitis: stop smoking.

Cigarette smoking can also cause shortness of breath. It can cause disease of the bronchi which leads to an obstruction in the flow of air, a condition sometimes called **chronic obstructive lung disease**. It can also destroy the spongy or alveolar part of the lung, leaving large holes and a loss of elasticity. This is called **emphysema**. The combination of these two diseases can be disastrous and unfortunately can never be made better. Additional damage may, however, be prevented by stopping cigarette smoking.

SHORTNESS OF BREATH

Prevent shortness of breath: stop smoking.

Asthma is another respiratory disease that mainly affects the bronchial tubes. Cigarette smoke is not the main culprit responsible, although it can make the symptoms worse. Inflammation of the internal lining of the windpipes takes place because of allergy to pollens (see Chapter 7), house dust mites, etc., or because of infection with viruses and other germs. This inflammation leads to swelling and thus obstruction of the bronchial tubes (similar to nasal obstruction during a cold). In asthma, the inside lining and the muscles surrounding the bronchial tubes also tend to hyperreact by narrowing to stimuli such as cold, dry air, dust and exercise. This combination of changes leads to a cough, with or without sputum production, and the narrowed bronchial tubes make a wheezing or musical noise. All the extra effort required to breathe makes the sufferer feel short of breath even at rest. Unlike cigarette-induced damage, these symptoms tend to develop quickly and then resolve themselves completely or partially until the next attack.

About 10 per cent of children have suffered these symptoms before the age of eleven. The symptoms may, however, develop much more insidiously, especially in older subjects, and may be mistaken for cigarette disease. Classically in asthma the symptoms are much more pronounced at night or in the early hours of the morning. In some patients these symptoms may only be noticed during colds, with exercise or after exposure to animals. There is often a family history of the condition and an adult may remember waking from sleep as a child with some of the above symptoms. (In days gone by and still in some countries, these childhood symptoms were called bronchitis.) Frequently the only clue in the past was that of a personal or family history of hay fever or itchy skin rashes such as eczema.

Asthma is frequently mild and easily treated but it is occasionally severe. Because of this it is important to recognise the symptoms for what they are and to consult a doctor who will confirm the diagnosis, assess the degree of impairment and consider ways of preventing or treating the asthma. Simple avoidance of obvious precipitating causes may be sufficient to stop attacks in patients. Most sufferers require treatment to reduce the inflammation and to relax the spasm in the bronchial muscle. This treatment is frequently given in the form of an inhaler which enables the smallest quantity of the drug to get to the region where it is most needed. Tablets and other preparations are also available. It is important to anticipate symptoms and try to prevent them rather than to postpone treatment until the symptoms are unbearable.

ASTHMA

If you have a cough, sputum or wheezy shortness of breath, you may have asthma.

Asthma is treatable.

Lung cancer is another disease which mainly affects the bronchial tubes. It initially appears as a swelling or ulcer in the lining of the bronchus. Because of this it usually gives rise to a cough and sputum which may or may not be blood-streaked. Blood in the sputum is more frequently caused by disease other than cancer and should always be investigated. The disease may cause pain, shortness of breath and weight loss. It is possible in most sufferers to ease discomfort and in a few to cure the disease by surgery if diagnosed early enough. The encouraging point to make is that the majority of cases can be prevented by never smoking. Too many farmers have died unnecessarily from this disease.

LUNG CANCER

Prevent lung cancer: stop smoking.

Stomach and Bowel Disease

Infection is the most common cause of bowel disease in worldwide terms. Thus cholera, salmonella, dysentry, amoebae, worms and viruses are frequent causes of ill health. All episodes of diarrhoea and other bowel disease must be properly investigated to enable a diagnosis to be made and thus proper preventive measures instituted. Stool samples are usually sent to a laboratory and blood samples are sometimes required. Unfortunately in some countries, acquired immune deficiency syndrome (see later) is becoming increasingly common as a cause of bowel infection in both urban and rural communities.

Many people, at some time in their lives, have had a little indigestion with fullness in the stomach. Flatulence (belching) and occasional lower chest or upper abdominal burning may also accompany this. The latter is due to the escape of acid from the stomach into the lower part of the oesophagus or gullet. Aspirin-like drugs may also irritate the stomach. These symptoms are usually reduced by the use of a simple antacid, reduction of alcohol intake and the avoidance of tobacco.

If, despite these commonsense measures, symptoms persist,

then it is possible that a **duodenal** or **gastric ulcer** has developed. When such an ulcer is present, food intake may exacerbate or ease abdominal discomfort, which may become sufficiently severe to waken the sufferer during the night. Vomiting and weight loss may also be a feature. If rapid bleeding is taking place, blood may appear in the vomitus, where it may look like coffee grounds. It may also appear black and tar-like in the faeces. With this degree of bleeding, the sufferer feels acutely weak. In less acute cases anaemia will be present. Ulcers can now be diagnosed with great accuracy and good treatment is available if given early before structural damage develops.

Some people, especially if overweight, have part of the stomach pushed into the chest and this is called a **hiatus hernia**. This allows the escape of acid into the lower oesophagus when lying down or stooping. If this is not corrected medically or surgically, the resulting damage can cause a narrowing or stricture to develop. This makes swallowing of solid food difficult. Occasionally **gastric cancer** is present and requires early diagnosis if therapy is to be effective. The symptoms may be similar to those of a gastric ulcer or there may be decreased appetite, a feeling of a full stomach after a small meal and weight loss. The sufferer is frequently over fifty years old.

Gall stones may also cause abdominal discomfort, which is often more marked after eating fatty foods. If a stone gets into the bile duct, it may prevent the passage of bile from the liver into the intestines. This leads to **jaundice** with dark, yellow-discoloured urine and pale or white faeces. There are, of course, many other causes of jaundice including infections, poisons, cancer, allergies and blood disorders.

The large bowel may become inflamed leading to diarrhoea or constipation. These forms of **colitis (ulcerative colitis** and **diverticulitis)** tend to send the sufferer to the doctor. It is important, however, to bring mild changes of bowel function to medical notice, especially if the pattern of a lifetime has changed without any obvious cause. The same applies to the appearance of blood in the faeces or stools. Although there are many simple explanations, changes in bowel habits and blood in faeces may represent the presence of early **bowel cancer** which can be satisfactorily removed.

Both stomach ulcers and bowel disease may be mistaken for a weakness due to anaemia. Tiredness should not be dismissed simply as something to be expected from getting older. A blood count is easily checked, and if abnormal will result in the relevant investigations getting carried out. Remember that **haemorrhoids** (piles), an obvious cause of bleeding, may be secondary to underlying bowel disease. New haemorrhoids or symptoms should be brought to the doctor's attention.

Urinary Tract

One of the most common problems suffered by men as they get older is that of obstruction to the outlet of the bladder by an **enlarged prostate gland**. This is usually obvious initially as some slowness in the urine stream. The sufferer may have to wait for the stream to begin and there may be dribbling or a feeling of urgency to micturate. Less frequently the symptom is acute pain from a sudden obstruction. The symptoms are usually present during the day but may cause trouble during the night with the passing of small amounts of urine at frequent intervals. If these symptoms persist, then the doctor can examine the prostate gland and arrange for investigation and possible surgery. Surgery is now frequently carried out by a small internal operation rather than through an incision in the lower abdomen.

Women develop urinary infections more frequently than men. They may also have stress incontinence (involuntary micturition when coughing, laughing or lifting weights), especially if they have had many pregnancies. This should be prevented by good obstetric care, and the condition can be treated.

In both men and women a burning sensation when passing urine or the presence of blood should always lead to a consultation with the doctor to exclude infection, stones or, less frequently, cancer.

Diabetes

Diabetes is a common illness which may develop early in life, but more frequently appears in later life. All diabetes is due to a relative shortage of the hormone insulin. In the younger variety of the disease, insulin injections are invariably required to keep the blood sugar levels in control. The adult variety is usually milder, and the problem is often caused by excess demands for insulin due to a high intake of food with resulting obesity.

Severe diabetics have symptoms that bring patients automatically to medical attention, and thus it is more important to be aware of the risk factors and early symptoms for late onset or adult diabetes.

In all patients with diabetes the blood sugar level is high because there is insufficient insulin to bring sugar into the cells of the body. Some of the sugar will spill out into urine and will bring water with it. Because of this, the patient will notice an increase in urinary frequency which may continue during the night. Unlike prostate gland trouble, the volumes of urine are large and therefore the patient becomes dry and thirsty, leading to exhaustion and collapse if intake does not match output of fluids. Continued deterioration results in the breath developing a sweat (ketone) smell. The sugar in the blood is associated with abnormalities of the blood fat levels; all the arterial diseases mentioned in previous sections may appear and there will be a reduction in eyesight.

Diagnosis of diabetes can be carried out simply by measuring the blood sugar level after fasting or several hours after a standard glucose meal. Treatment depends on the variety of the illness. Those who are acutely and seriously unwell require fluids and insulin. Young patients generally need to remain in insulin therapy for the rest of their lives. Older patients, especially if initially overweight, often require only good attention to diet and less frequently tablet treatment.

Tuberculosis

In worldwide terms this is still a major cause of disease and death. It is usually caused by a germ called *Mycobacterium tuberculosis hominis*, which is a very slow-growing germ compared with most bacteria. There are also other human and animal varieties and the latter are dealt with in Chapter 8. In the Western world the number of patients with the disease has greatly diminished. For instance, in Northern Ireland where there are 1.5 million people, there were 3,000 new cases of pulmonary tuberculosis each year in the early 1950s in contrast to between 50 and 120 new cases yearly now.

A social stigma not limited by social class or education is still attached to the disease. It was thought in the past that the disease ran in families because of inherited characteristics. It is more likely that the slow-growing germ passed from parent to child in the household because of close contact, especially in overcrowded homes.

Because it normally affects the lungs, the main symptoms are those of a cough and sputum, the latter occasionally being blood-stained, particularly in advanced disease. Along with the chest symptoms there may be weight loss, sweating and prolonged fever. Other parts of the body may be affected. The diagnosis is usually made by use of a combination of skin testing, x-rays and a culture of the germ from the sputum.

The disease is now completely curable by drugs taken by mouth, and hospital admission is rarely required. The assumption is some-times made that blood in the sputum means lung cancer, and consultation is delayed, with unfortunate consequences. AIDS has slowed up or even reversed the downward trend in the incidence of the disease.

TUBERCULOSIS

Chronic ill health, cough and sputum may indicate tuberculosis, which is treatable.

Gynaecological Problems

Perhaps the most common gynaecological problem in women is that of excessive menstrual bleeding. This may be due to heavy bleeding with clots, prolonged bleeding or too frequent periods. Inevitably, if blood loss exceeds its replacement by the bone marrow, anaemia will result. This is frequently hastened by poor dietary intake of iron or frequent childbearing. At an early stage tiredness may be the only symptom, but later pallor becomes obvious (initially seen in the tongue or inside lining of the eyelids), along with subsequent shortness of breath.

Vaginal bleeding between periods, especially in women over thirty years of age, should always be investigated to exclude infection, ulcers and cancer of the cervix and uterus.

Cervical cancer is common but in many Western countries it is less frequent than cancer of the lung. It may cause vaginal discharge or bleeding. Many countries now have cervical smear screening programs to try to diagnose the disease while it is still superficial and curable.

Secretion is normal but an excessive vaginal discharge can cause

41

distress due to the staining of clothes, its unpleasant odour or itching. Although a serious underlying cause cannot be excluded, most discharges can be cured.

GYNAECOLOGICAL DISEASES

Many gynaecological diseases are preventable and curable.

Join disease screening programs.

Breast Cancer

Breast cancer is the most common cancer affecting women. Not all lumps in the breast are due to cancer. A woman should examine her breasts every few months, using the palm of the hand to cover each part in sequence. If she has not been shown the correct method, this can be demonstrated by a doctor or nurse. If there is a health screening program in the neighbourhood, it should be attended.

Acquired Immune Deficiency Syndrome (AIDS)

This disease is caused by one of the human immunodeficiency viruses (HIV). Adults are usually infected through having sexual intercourse with an infected partner. In the Western world this has been predominantly through homosexual practice, but it is becoming increasingly a heterosexual problem as it is in Africa. Infection is also acquired from infected blood given for medical reasons or on needles shared by drug users. A foetus can be infected in the womb.

The virus infects the white blood corpuscles that are responsible for the defence of the body against infection. Blood tests can show, at an early stage, whether the patient has the virus. It is some years later before the disease becomes apparent to the patient or doctor. There may be swollen glands, feverishness and a lack of wellbeing. Later the patient becomes infected by germs that do not usually infect healthy people. AIDS patients frequently develop pneumonia, diarrhoea or meningitis. They may also get rare forms of cancer. Eventually they lose weight and die. Some treatments now available seem to delay the onset of the severe disease and can treat the infections. Vaccines are currently under study.

We are all, at the end of the day, responsible for the health of ourselves and our families. The above descriptions of diseases are brief and intended to stimulate interest in the types of ill health that are prevalent anywhere. It is only by knowing the facts about such diseases that you can learn how to avoid them or to keep the complications to a minimum.

5

Dusts, Sprays and Gases

The first section of this chapter deals with some general points about dusts, sprays and gases present on farms. Diseases caused by them are described in subsequent chapters.

Gases and small particles of solid or liquid suspended in the air all have the potential to be harmful.

Normally the air contains nitrogen, oxygen, carbon dioxide, water vapour and minute concentrations of rare gases. Unfortunately, in the modern world, this type of 'pure' air is probably only present over the middle of the Pacific Ocean, the remainder having been contaminated. Much is known about the harmful effects of many of these pollutants; however, it often takes years before scientists discover a link between exposure and the development of a disease. It would therefore be prudent to keep all unnecessary inhalants to a minimum. (See Appendix 3 for protective masks and helmets.)

Since these substances are in the air we breathe, they can cause damage to the bronchial tubes and lungs and subsequent disease. It must be remembered that many of the gases and particles will come into contact with the skin and may be absorbed through it into the blood. Likewise some will get into food, be swallowed, enter the intestines and be absorbed into the blood which carries the gases and particles to all the organs of the body. Different chemicals cause disease in different parts of the body.

In general, gases and fine particles enter the lungs while larger particles settle on the skin or food. Liquids are more likely to be absorbed through the skin than dusts, although dusts may dissolve in sweat, water or other liquids and thus be absorbed.

DUSTS

Dusts are small solid particles. The size, shape and density of the particle determines how quickly it will settle out from the air onto the ground. Small particles are more likely to be dispersed by the wind and thus come into contact with vegetation, animals or humans.

Dust particles small enough to be inhaled and yet still relatively large (10–100 microns in diameter) will nearly all be removed by the lining of the nose and throat. When the nose is subsequently blown or the throat cleared, the dust particles will be expelled. This is the natural mechanism for protecting the lungs from foreign substances. Sneezing may also help to dislodge mucus or dust. Smaller particles (5–10 microns in diameter) tend to be deposited upon the lining of the bronchial tubes and may cause a cough with sputum production and occasional swelling of the lining which may cause asthma. Even smaller particles (1–5 microns in diameter) may reach the spongy alveolar part of the lung where they can cause inflammation or scarring as in farmer's lung or asbestosis. Still smaller particles (less than 1 micron) tend to move in and out of the lung with the air but these do not settle.

Examples of particle sizes

Sand grains	20–2000 microns
Cement dust	4–100 microns
Pollens	10–100 microns
Mouldy hay spores	0.6–2.5 microns

The following are examples of some of the dusts to which the farmer may be exposed:

Dry soil dust   This contains small particles of sand or silica with variable amounts of organic content.

Germ spores and infectious particles   These can be present in the ground, in dust or on mouldy vegetable material. Animal skins or their carcasses may be contaminated, as may the products of abortion, new-born animals or their placentae.

Asbestos   This may be inhaled when cutting asbestos sheeting and in some countries such as Turkey and Romania it may be present in the soil.

Lime   This contains calcium carbonates and may also contain some silica.

Fertilisers   These are rich in the mineral or organic nitrates which encourage growth with phosphates and potassium. They may contain other substances such as talc to promote running of the material and to act as an anti-caking agent for powders and granules.

Weedkillers   These may be selective as to vegetation they kill or they may be generally lethal.

Pesticides   Like herbicides these may be of a selective or non-selective type.

Spores, mites and allergic particles   These may be released from normally stored material such as grain and hay or from mouldy vegetable matter.

Some of the diseases caused by the above dusts are described in detail later.

MISTS AND SPRAYS

Mists are very small liquid particles that tend to remain suspended in the air. Sprays are larger droplets that tend to drop to the ground. The same general principles apply to mists and sprays as to dusts. Smaller, less dense particles tend to be easily blown off course onto areas for which the spray or mist is not intended, and liquid particles may be inhaled or get onto the skin and into the eyes or the mouth.

Because of this it is very important to read the instructions that come with liquid preparations and to ensure that the spraying apparatus used produces the correct particle size. Altering the concentration of the liquid by dilution with water or increasing the operating pressures may change the particle size to that which may be inhaled. Masks (Appendix 3) may be needed to keep toxic substances from entering the body this way. It must also be remembered that chemicals from sprays and mists may penetrate the skin, especially if it is moist or cut, and thus protective clothing should be used if indicated by the manufacturer.

Sprays and mists must be used when applying liquid weedkillers or herbicides, but they can also be produced inadvertently by high-pressure water hoses when washing down milking parlours or other areas on the farm. This may lead to the inhalation of infectious particles or skin contamination by them.

GASES

Not all potentially lethal or dangerous gases can be seen or smelt. To enable adequate precautions to be taken, it is therefore essential to know the areas and processes on a farm from which these gases might originate.

Tractors, or other fuel-driven apparatus, produce carbon dioxide and carbon monoxide, amongst other gases, in the exhaust fumes. In enclosed spaces the levels of these can build up. **Carbon dioxide** causes drowsiness and headache but needs to be at quite a high level to create these problems. **Carbon monoxide** is exceedingly dangerous even at low levels.

Vertical silos may be dangerous due to the production of gases from the breakdown of grain and silage. The heavy gases **carbon dioxide** and **nitrogen dioxide** accumulate in the hollows in enclosed spaces at the tops of silos, and their combined effects in these areas may be lethal.

Slurry pits produce **hydrogen sulphide**, a gas which is dangerous to both lungs and brain. **Methane** gas is also produced and this can act as a fuel which is capable of explosion.

Pig houses may contain a high concentration of the irritant gas **ammonia**.

Most toxic gases can be reduced in concentration or removed by good ventilation, especially when access is required. Adequate time must be allowed for the gas to escape, and it should be noted that heavy gases will not escape unless a draught is going through the container or building. The general precautions given in the section on accidents must be adhered to.

In the following chapters, the risks from some of the individual gases, sprays, mists and dusts are given. They are divided into substances which cause disease by being irritants or poisons and those causing allergic or immunological diseases. Infectious risks are dealt with in Chapter 8.

47

6

Dangerous Irritants and Poisons

This chapter deals with substances that can damage the body by some method that is not principally related to allergy or infection. The effects can be temporary or permanent.

Gases

NITROGEN OXIDES

Many oxides of nitrogen are formed in vertical silos, with nitrous oxide at the bottom of the silo and nitrogen dioxide at the top. The latter is a heavy, brown-coloured gas and tends to fall into and accumulate in hollows of grain in the middle or around the edges at the top of the silo. Not only is nitrogen dioxide inherently dangerous by itself but, together with other less toxic gases such as carbon dioxide, it displaces the oxygen from the hollows. Thus a farmer working at the top of a silo or in the feed room near an open access door will be starved of oxygen and may lose consciousness quickly and without warning.

Small concentrations of nitrogen dioxide (NO_2) from as low as five parts per million may cause bronchial and alveolar lung disease some hours after exposure. At sixty parts per million, throat irritation is noticed and at one hundred parts per million coughing will be common. With higher concentrations, damaged blood vessels in the lung will begin to leak from three to thirty hours after exposure, even though the farmer is removed from the gas. This results in a cough, wheezing and shortness of breath. Bloody, frothy sputum is coughed up and in more severe cases the farmer will collapse and may die. If the farmer survives this severe acute damage, recovery is usually complete.

HYDROGEN SULPHIDE

This gas usually develops in slurry tanks. Although it has the strong smell of rotten eggs this is usually ignored because the farmer or his children expect such a smell from slurry. Like other gases, hydrogen sulphide (H_2S) may kill by simply reducing the oxygen in the air above the slurry. At a concentration of seven hundred parts per million this gas can rapidly cause death by its direct action upon the part of the brain which makes us breathe automatically. Smaller concentrations can cause effects similar to those of nitrogen dioxide upon the bronchial tubes and lungs.

CARBON MONOXIDE

This gas from exhaust fumes or town gas is odourless and colourless. It binds onto the red blood corpuscle haemoglobin with a binding power two hundred times greater than that of oxygen and harms the body by depriving it of this necessary gas. Unfortunately, despite the lack of oxygen, the skin of an individual affected looks red and not blue and the unwary observer may not recognise the danger. A low concentration of around fifty parts per million over an eight-hour working day may lead to such effects as agitation, headache and occasionally confusion. As the concentration builds up, nausea and vomiting may result or the farmer may lose consciousness.

AMMONIA

This is a very irritating gas which tends to make the eyes and nose run and leads to coughing, sputum and occasionally shortness of breath. Patients with asthma tend to be more susceptible to this form of irritation, which may precipitate an attack. High concentrations have effects similar to those of nitrogen dioxide upon the bronchial tubes and lungs.

Dusts

ASBESTOS

This substance is sometimes present in the soil and thus in some countries such as areas of Turkey, Romania and Finland, exposure may take place during ploughing, especially in dry weather. Exposure may also take place when sawing asbestos boards or sheeting for roofs. To keep exposure to a minimum, such a procedure should be done out-of-doors. If it has to be done indoors, then a properly fitted and effective mask should be used.

Small concentrations of asbestos dust may result in thickening of the lining (pleura) around the outside of the lung and inside the chest wall. Less frequently a rare cancer (a mesothelioma) of this lining develops. Much higher and usually industrial concentrations of asbestos dust are required to produce significant scarring (fibrosis) of the alveolae of the lungs, a condition known as **asbestosis**. In asbestosis, breathing becomes difficult due to stiffness of the lungs and ultimately the heart fails. A bronchial cancer may develop. There is no cure for any of these asbestos diseases and thus prevention is essential.

LIME

Most of the particles of lime are fortunately too large to be inhaled. They may, however, be locally irritating to the mouth, nose and skin. The eyes are in special danger from the alkaline effects of the dust, and particles may have to be removed physically if they do not wash out easily with clean water. This irritation results in a red, watery eye. If the central part of the eye (the cornea) is involved, scarring may result with subsequent diminution of vision.

Sprays

This section on irritant and toxic sprays deals mainly with pesticides and herbicides. The reader will have to learn the laws and regulations that apply to the use of these substances in his or her own country. For instance, in the UK the Control of Pesticides Regulations specify those who are allowed to use these chemicals and state the conditions of sale and use. A certificate of competence is required by the responsible person and all users must be adequately instructed and supervised. A certificate of competence is now required by anyone born after 31 December 1964 before work can be done unsupervised. The areas covered by this training include the following:

knowledge of regulations	calibration of machinery
when to use pesticides	safe operation of machinery
appropriate pesticides	safe storage and disposal
labels and codes	response to poisoning
health and environmental risks	response to contamination
protective clothing	accidental spillage
engineering controls	record keeping
monitoring exposure	health surveillance

The Health and Safety Executive of the Ministry of Agriculture, Fisheries and Food for the UK also produces a list of approved pesticides and herbicides. It is illegal to use any substance in the UK that is not currently on this list. It is therefore important that the farmer obtains information about *current* lists before each new season. This is another good stimulus for not storing chemicals for prolonged periods.

ORGANOCHLORINE PESTICIDES

These substances include the following:

aldrin	DDT	lindane
endrin	kelthane	chlorinated terpines
chlordane	methoxychlor	toxaphene
heptachlor	chlorbenzylate	

They are used much less than previously, and many have been banned. They tend to be absorbed and stored in fatty tissues in the body and hence tend to accumulate over a long period of time with resulting chronic toxicity.

Acute poisoning is not common unless large quantities are drunk accidentally or absorbed through the skin. This will result in nervous excitement, shaking and convulsions. Nausea and vomiting may also occur. If the patient survives the early effects, liver and renal failure then develop. Chronic poisoning from skin absorption affects the same parts of the body with the addition of lung disease. There is also concern about the possibility of cancer developing as a result of chronic exposure.

All these substances are also skin irritants, and local damage may

51

result, with the additional risk of the entry of infectious germs into the skin.

Hexachlorbenzene, which is used as a seed dressing, may, with chronic exposure, lead to severe blistering of the skin, especially in areas exposed to sunlight. The blisters may heal with scarring. Many other organs in the body may also be affected by absorption through the skin with resultant disease in the bones, joints, liver and thyroid gland.

ORGANOPHOSPHATE PESTICIDES

They include the following:

parathion	diazinion
methyl-parathion	phosmet
dichlorvos	dimeton
trichlorfon	and others

These pesticides affect man in a similar way to the nerve gases used in some wars and, like them, are very well absorbed through the skin. They prevent the natural breakdown of acetylcholine, a natural chemical in the body responsible for nerve transmission.

As a result of the prolonged, high concentrations of acetylcholine, nerves, glands and muscles continue their action for longer than is necessary. In acute poisoning the pupils of the eyes are small, and salivation and sweating take place. This may proceed to nausea, vomiting and abdominal pain. The muscles twitch and go into prolonged spasm with subsequent exhaustion. Chest tightness, coughing and wheezing affect respiration, which ultimately may stop due to brain failure and coma. If the patient survives, the symptoms may persist for days or weeks.

CARBAMATE PESTICIDES

The effects of these pesticides tend to be similar to those of the organophosphates but they are of shorter duration. They include the following:

asulam	methiocarb
barbanc	terbutol

PHENOXY COMPOUND HERBICIDES

These are hormonal, selective weedkillers:

2,4-D   fenoprop
2,4-DB   mecoprop
dichloroprop

They are strong local irritants if they get on the skin or into the eyes or mouth. Large concentrations, if absorbed, may cause salivation, vomiting and abdominal pain. Muscle weakness may, if severe, lead to coma. There is also concern about foetal abnormality and the production of cancer.

DIPYRIDYLIUM HERBICIDES

The most commonly used and probably the most toxic of this group is Paraquat. It can be absorbed into the body from external as well as internal contact. Blistering of the skin and mouth results. After a lapse of usually several days, severe lung and kidney disease develops and this almost invariably leads to death.

Even a small quantity of the chemical is fatal for most individuals. Drinking Paraquat may be accidental or by design but in either case the illness is distressing for the patient, the family and the hospital staff.

TRIAZINE, UREA COMPOUND AND ANILIDE HERBICIDES

These may cause irritation to whichever part of the body they come into contact with and may also affect the respiratory tract.

It is obvious from the above that contact with these substances may have disastrous consequences. It is therefore essential that good codes of working practice are followed when any of these substances are being used. The advice in the manufacturer's literature must always be adhered to and, if in doubt, contact the Ministry of Agriculture, Fisheries and Food or other government agency. Because of the cost involved, there is frequently a temptation to use the cheapest preparation available. This is acceptable if it is government approved. Black market preparations may not be produced to the same standards and can be dangerous.

Correct protective clothing must always be used even for small jobs and emergencies. It must also be worn during the preparation stages and not just when the spraying begins. Likewise, protective clothing should be worn in warm weather since sweating may enhance skin absorption.

Helpers, family and other onlookers must either be excluded or properly protected. Previous advice about the storage of chemicals should be adhered to.

Treat all chemicals with caution.

7

Diseases Due to Allergy or Immunity

In extrinsic allergic alveolitis, asthma and other allergies, the illness is not caused by infection, even though bacteria and fungi may be involved in the general process. Likewise the substances causing damage are not direct irritants or poisons. In each of the illnesses described in this chapter, the body has to develop an antibody reaction against an inhaled substance before the illness can begin.

Extrinsic Allergic Alveolitis

These diseases are caused by the inhalation of organic material in dust or droplet form. The inhaled particles, which may not be visible, are ingested by special body cells (macrophages) which digest the particles and send messages to the antibody-producing cells (lymphocytes and plasma cells). These cells produce a special pro-tein—an antibody—which attaches only to the inhaled substance—an antigen—that induced its formation. In infectious diseases antibodies so formed will damage the germs, thus enabling the body's defences to kill the germ and cure the illness. However, in extrinsic allergic alveolitis, exposure to the organic material is usually constant or intermittent. If a farmer who already has antibodies present in his blood and lung tissues inhales foreign material, the antigen and antibody interact and behave like a detonator to start a chain reac-tion with other body proteins and cells. This ultimately results in the formation of substances capable of causing acute inflammation. Because of the small size of the dust particles, the interaction with the antibody usually takes place in the alveoli. This reaction in the alveolar walls causes dilated, leaky blood capillaries out of which fluid and white blood corpuscles pour. This is in fact a non-infectious pneumonia or alveolitis.

Examples of these diseases are bird-fancier's lung, mushroom-worker's lung and farmer's lung. All these illnesses are similar in nature and the example of farmer's lung will be used below. It should, however, be remembered that all organic material, vegetable or animal, if inhaled can theoretically produce extrinsic allergic alveolitis. The best principle must therefore be to always keep dust exposure to a minimum and to wear a mask if necessary.

Some Causes of Extrinsic Allergic Alveolitis

name of disease	cause
farmer's lung	mouldy hay, straw or grain
bagassosis	mouldy sugar cane fibre
mushroom-worker's lung	mouldy compost
malt-worker's lung	mould on germinating grain
sequoiosis	mould in redwood bark
suberosis	mould in cork bark
cheese-worker's lung	mould on cheese
bird-fancier's lung	bird feathers and droppings
fishmeal lung	fishmeal dust
weevil alveolitis	wheat weevil
vineyard-sprayer's lung	Bordeaux mixture
insecticide lung	pyrethrum

Farmer's Lung

This is acquired by the inhalation of spores from thermophilic or heat-liking moulds (e.g. *Micropolyspora faenae*). These moulds need moisture and heat to enable them to germinate and grow. If vegetable materials such as hay, straw and grain are stored damp (greater than 30 per cent moisture content), they are ideal culture media for the bacteria to grow in and this causes the temperature to rise. When the right temperature is reached (30–65 degrees centigrade), usually after several months of storage, the thermophilic moulds start to germinate. When hay is subsequently disturbed by forking or shaking out, millions of spores are released into the atmosphere and will be available for the farmer to breathe in. The number of spores inhaled depends upon a number of factors. Nasal breathing will reduce some of the spores by trapping them in the nose. However, because of their small size, most will proceed into the lungs. Hard work will increase the quantity of spores arriving at the alveoli because of the increased volume of air breathed. If the hay or grain is very mouldy, the concentration in the air will be greater and only a short time in the area may be sufficient to cause disease.

Many people, especially those with asthma, will develop a cough at the time they are working in the dust due to the irritation of the bronchial lining. Although important by itself as a symptom of irritation, this is not an indicator for the presence of farmer's lung. Classically a farmer with the disease will notice the symptoms developing several hours after working in the dust because it takes time for the reaction with the antibody to take place. The farmer may at this stage develop a dry cough and feel that he has influenza or flu-like symptoms such as chills, aches and pains. Shortness of breath will be present if the disease is more marked. Obviously if these symptoms are acute, the farmer may recognise the relationship to dust exposure or he may consult his doctor. If the symptoms are not acute they may not be linked to dust exposure and there will be continued contact with the spores over the winter months. If this happens, inflammation in the lung becomes chronic and leads to scar tissue developing. This causes irreversible damage resulting in permanent shortness of breath. Weight loss and blood in the sputum may develop and may be misdiagnosed as lung cancer or tuberculosis.

Diagnosis of extrinsic allergic alveolitis is made by looking at the history of exposure to dusts and some of the symptoms mentioned above. With the suspicion of this disease in mind the doctor may listen to the chest and hear crackling noises. Breathing tests and chest x-rays may demonstrate changes compatible with alveolitis and will show the extent of the disease. A blood sample will be taken to demonstrate the presence of antibodies produced against the spores or extracts of mouldy hay. Occasionally antibodies are found in the blood even when the farmer has no clinical symptoms. This finding should not be ignored since it indicates either early disease or a susceptibility to the disease in the right circumstances.

If the disease is present, then the farmer must be removed immediately from all contact with the responsible dust. If the disease is mild this will be sufficient to cure the patient. More acutely severe symptoms may require the use of cortisone drugs (steroids) to suppress the inflammation in the lungs. Oxygen is given if the blood levels are low. If scarring has developed, although the above still applies, the final result will not be as satisfactory because the scar tissue remains and may cause continued shortness of breath.

The answer to farmer's lung must lie in prevention. If, for financial reasons, poor quality or obviously mouldy hay, straw or grain has to be used it should be disturbed as little as possible. If the bales of hay or straw have to be opened, this should be done out-of-doors and downwind from the farmer. Good masks (see Appendix 2) should be worn if there is to be prolonged contact and the ventilation should be sufficient to dilute the number of spores in the air. The use of silage is not associated with farmer's lung, so conversion to this method of feeding will prevent development of the disease.

If a diagnosis of farmer's lung is made, the farmer must choose between changing his type of farming or instituting the preventive methods discussed above. If the latter course is taken, then some element of risk continues and regular checks must be made so that the degree of risk in the particular circumstances can be constantly assessed.

Asthma

Farmers are just as likely to suffer from common asthma as other members of the community, and this aspect is dealt with in Chapter 4.

Farmers will, however, have additional problems because of the nature of their work. In season, they will be near pollinating grasses, plants and trees with resulting high concentrations of pollen grains which may lead to the development of an allergy. There are allergic substances in the dusts from hay, straw, grain and other vegetable products. Likewise animals are all capable of causing allergic reactions and asthma. Both the early morning start and the cold air will tend to precipitate asthma on winter mornings and this likelihood will increase if the work is strenuous, either in the farmyard or in the fields. All dusts and sprays, even if they do not cause allergy, may irritate the bronchial tubes and lead to asthmatic symptoms in susceptible farmers.

It is interesting that in Northern Ireland we found 40 per cent of farmers questioned gave a history of wheezing at some time in their lives, although only 4 per cent had been told they had asthma. When working with hay, 24 per cent had a cough and 14 per cent wheezed. These results suggest that asthma is being under-diagnosed and hence under-treated in the farming community.

Other Allergic Conditions

Allergy occurs most frequently in the eyes, nose and skin, although it can affect other parts of the body.

EYES

Allergy in the eyes usually shows itself as redness of the part that is usually white and is associated with itching, watering and discharge. Swelling of the eyelids may also occur. The most common cause is probably an allergy to pollens or other airborne dust, but it may also result from rubbing allergic substances into the eyes. occasionally foods or drugs are responsible.

NOSE

Allergy in the nose leads to an increase in nasal secretion and swelling of the nasal lining. This results in nasal discharge which may be clear or coloured and which pours from the front or the back of the nose. Itching or sneezing may be present and there may be difficulty in breathing through the nose. If the swelling becomes more severe, the nose may become completely blocked with resulting loss of smell and discomfort in the nose or surrounding sinuses. Occasionally a grape-like swelling of the lining develops (a polyp). The condition may be acute with long, symptom-free intervals or it may be perennial. Its causes are similar to those of allergic conjunctivitis and asthma.

SKIN

Allergic skin rashes may be divided into several main groups: urticaria, atopic eczema and contact dermatitis.

Urticaria is a very itchy, pale, raised rash which is often surrounded by redness and looks like a common nettle rash or small insect bite.

59

It may appear as many small lesions, or a large confluent area, such as the lip or an eyelid, can be swollen. It is usually potentially dangerous only if the tongue or throat is involved. It can be caused by many allergic substances and, like all the skin allergies, is more likely to be caused by direct contact or by food (or food additives) and drugs. Very occasionally it is related to the deficiency of a special inhibitor of allergic reactions in the blood.

Atopic eczema is more common in patients who have a tendency to develop asthma or allergic rhinitis. It is an itchy, red patch of skin which may become moist from scratching or, in a more chronic stage, the skin may be thickened and scaly. It usually attacks the skin at the creases of the wrists, elbows and behind the knees but can affect other areas as well. It is caused by the same things as asthma, with perhaps a greater element from food allergy and direct contact.

Contact dermatitis is an inflamed and sometimes itchy area of the skin which often has blisters or raised areas, especially at the edge of the affected area, and if chronic the skin becomes thickened and cracked. It is most common on the hands, forearms and face. Although it may be caused by the same substances that cause urticaria, the white blood corpuscles tend to play a greater part than the antibodies in the formation of this condition and it may be initiated by such simple chemicals as cement and nickel.

All these conditions may be made worse by drying of the skin's surface in windy, cold conditions or by soaking in water. Ill-fitting clothes may also make things worse by irritation or the retention of the allergic substances. Once the condition is recognised, obvious allergic substances can be avoided and, if this does not prevent or remove the rash, medicated creams can be used to reduce the inflammation and treat secondary infection if necessary. The application of a good conditioning cream and subsequent washing after contact with harmful substances will help to prevent these diseases from developing.

WASP AND BEE STINGS

These may produce localised urticarial reactions or much more generalised ones. In some bee keepers there may be no reaction due to desensitisation from previous bites. Some people may have localised urticarial reactions with pain, more marked if the sting is in an area where the skin is tight, e.g. on the finger. In other areas the swelling may be sudden and massive such as around the eyes or the mouth. Occasionally a more severe generalised reaction takes place with resulting asthma or even collapse. Such emergencies are rare, but if reactions have been severe, a doctor should be consulted.

8

Zoonotic Infections

Zoonotic infections are infectious diseases transmitted from animals to man. Many of these illnesses are of a mild nature, and the farmer may think that he simply has a chill or flu-like symptoms. There are several reasons why it is important to try to make a definite diagnosis. If cases are not diagnosed and reported, the public health authorities will not be aware of the extent of a disease in the community, and thus epidemic proportions may be reached before preventive measures are taken. Reporting infectious disease is important because it helps alert the doctor to a type of illness prevalent in the community, and hence a diagnosis may be made more quickly and treatment started. The reporting of disease can also be of benefit to the general community. For instance rubella infection (German measles) is reported, not because it is dangerous to the patient, but because it is highly damaging to the unborn child. The same could be said in farming about toxoplasmosis.

Even if a single disease is relatively mild, repeated episodes of one or more diseases may lead to chronic ill-health with all the implications for the farmer in terms of efficiency. Many zoonotic infections are potentially very serious and should be diagnosed as soon as possible to enable treatment to be given before complications result. Failure to make a diagnosis and to report the illness may be dangerous not only to the individual farmer but also to his family and the community at large.

Only when the size of a problem is recognised will the effort and funds be put into research and into the medical and paramedical services needed to prevent and treat the diseases. Too often commercial interests related to farm animals or media pressures in the apparent interests of the consumer take precedence over the consideration of the effects of disease upon the producer. It is in the farmer's own interest to try to put his problems into perspective.

Infection in animals, including man, is caused by germs and

parasites that are pathogenic (capable of producing disease). These pathogens may be specific to a particular animal or species or they may infect many species. Zoonotic infections are part of the latter group. The infectious agents range in size from very large worms to very small viruses capable of multiplying only if they are inside individual cells of the body. A pathogen may cause disease in the form in which it gets into the animal, e.g. salmonella, or it may enter in a different form before eventually maturing into the recognised infection, e.g. many worm infections. Because of the ability of most animals to protect themselves against infection, it is usually necessary for a high concentration of pathogens to be present at any single exposure before disease develops. In susceptible individuals, however, most exposures are unfortunately capable of causing disease.

Depending upon the individual germ, disease may result from direct contact with the skin and eyes or with any of the internal body linings after swallowing or inhalation. Infection is more likely to occur if the skin is damaged by trauma. Excessive wetting of the skin by sweat or immersion in water may allow germs to penetrate more easily. This will also happen if the skin is cracked after drying. Sometimes a carrier (vector) of the germ, such as a tick, is required to inject it through otherwise healthy skin. Other infections are acquired by the inhalation of infected droplets from the air. These can be produced during talking or coughing and by mists and sprays created by hoses and equipment. Dust from infected animals and soil will cause the same problems. The particles get into the nose, throat and lungs causing colds, sore throats and bronchitis or pneumonia.

The disease may remain localised in the place it initially penetrates and infects. The germs may, however, get carried in the lymph to the lymph glands which become swollen. They may also get into the bloodstream (septicaemia) and be carried to any part of the body.

The body reacts to most infections by a rise in temperature although, especially in the elderly, severe disease may be present without a fever. There are a number of symptoms that can occur with a rise in body temperature irrespective of the cause, such as shivering, aches and pains, headaches and sweating. These flu-like symptoms are often mistakenly thought to be due to influenza or even malaria and their true cause not investigated.

Details of treatment are not given for the diseases in this chapter since self-medication is not advisable without advice from your doctor. There are, however, a number of general points to be remembered:

63

1. General care The patient must be kept well hydrated by encouraging fluid intake. Intravenous drips are occasionally needed. Fever symptoms can be eased by simple drugs and tepid sponging, and pain can usually be relieved. If pus is present, the patient will rarely feel better until it has been released. Intensive care in hospital may be needed to keep the respiration, circulation and kidneys functioning while awaiting the outcome of some severe infections.

2. Viral diseases Few zoonotic virus infections can be cured with drugs, and general care is the only treatment available while awaiting the natural outcome. Prevention is therefore essential.

3. Bacterial diseases Most can be treated with antibiotics. To select the best drug, the diagnosis should be made accurately.

4. Parasitic diseases Although most can be treated with drugs, some of the injectable drugs can be very toxic. Surgery is occasionally necessary. Residual damage, even after the parasite has been killed, may be severe.

A number of zoonotic diseases are now described briefly. I have also included descriptions of a few diseases that are not truly zoonotic but are frequently seen in farmers. They are placed in alphabetical order within groups according to the infectious agent, i.e. viruses, bacteria, fungi, worms and flukes. The tables at the end of the chapter give the diseases that can be acquired from individual animals. They also list the different parts of the body that are affected.

Some infectious diseases in both man and animals must, by law, be reported to the relevant government department. The list of notifiable diseases in the UK appears in Appendix 2.

Viral Infection

COWPOX

This is caused by a pox virus which is similar to the smallpox virus. It is probably acquired from cows but may also be carried by cats and small rodents. The cows may have a vesicular rash on the udder and teats. The infection is transmitted from the teats to scratches on the milker's hands. After three to seven days' incubation a blister or vesicle appears which becomes pustular, and there may be swelling of the local lymph glands.

ENCEPHALITIS (ST. LOUIS, CENTRAL EUROPEAN, ETC.)

This group of central nervous system (brain) infections is caused by the Togaviridae. They are transmitted to man by ticks and mosquitos from a variety of animals (horses, goats, birds, etc.). Inflammation of the brain causes altered consciousness and paralysis.

HAEMORRHAGIC FEVER WITH RENAL SYNDROME (HANTAN FEVER)

This is caused by the Hantanvirus (a Bunyaviridae) and is carried by rats (*Rattus rattus*, *Rattus norvegicus*), field mice (*Apodemus agrarius*), wood mice (*Apodemus sylvaticus*) and the bank vole (*Clethrionomys glareolus*). It has also been found in the house mouse (*Mus musculus*). These animals excrete the virus in urine and saliva but do not suffer from the disease themselves. Man acquires it by contact with infected water or secretions. The incubation period is one to five weeks. In the Western world there is often a mild, influenza-like illness; in the East there may be severe prostration with high temperature, skin haemorrhages and blood and protein in the urine.

HAND, FOOT AND MOUTH DISEASE

This is caused by one of the Picornaviridae. Like most viruses it survives longer outside the body. It is a disease of cloven-hoofed animals which develop painful vesicles on the feet and both in and around the mouth. They usually drool saliva. It is acquired by man from the infected secretions, and after an incubation period of varying length, vesicles appear around the mouth and on the hands and feet. However, it is frequently asymptomatic (without symptoms), and it may be diagnosed only by testing the blood for antibodies to the disease.

LOUPING ILL

This is caused by one of the Flaviviridae and is spread by tick bites. The virus infects sheep, deer and ground-living birds in which brain infection leads to incoordination and eventual paralysis. Man only occasionally acquires the infection, either by tick bites or possibly by the inhalation of infected dust. After an incubation period of four to seven days there is an influenza-like illness followed by meningitis or nervous changes similar to those which occur with encephalitis, giving alteration of consciousness and paralysis.

A similar, recently discovered viral illness in cattle and possibly other animals is **bovine spongiform encephalopathy (BSE)**. It appears to have been spread by the introduction of infected brain tissue into the food chain. Intra-uterine transmission to calves may also occur. Infected animals lose coordination, stagger and become paralysed. There is no good evidence that it is transmitted to man but two diseases exist which follow a similar pattern. **Kuru** occurs in tribes in New Guinea and is probably caused by their eating infected human brain tissue. **Jacob-Creutzfeldt disease** is a rare disease in the Western world which may be transmitted by contact with brain tissue. The agent responsible for this condition is known to be very resistant to disinfectants. Both these human diseases cause staggering, involuntary movements and ultimately dementia. In view of the similarity, it is obviously important not to take risks with BSE-infected material. Since louping ill has been recognised for years and has not led to an obvious epidemic in man, then with care it is unlikely that one will occur with BSE.

LYMPHOCYTIC CHORIOMENINGITIS

This disease is caused by an arenavirus. It is carried by the house mouse which shows no symptoms but excretes the virus in its saliva, urine and faeces. It is acquired in man by contact with infected secretions or by bites. After one to two weeks' incubation an influenza-like illness may develop with orchitis (sore, inflamed, tender testes), meningitis or encephalitis.

NEWCASTLE DISEASE

This is caused by a paramyxovirus, which infects birds and causes respiratory disease. The infection in man is acquired by touch or inhalation of infected aerosols of dust, secretion or from live vaccine used to immunise birds. After a few days' incubation period, a painful conjunctivitis (red eye) develops with or without influenza-like symptoms.

ORF

The virus causing this condition is one of the Poxviridae. It is usually acquired from sheep or goats but may occasionally be acquired from cattle and even from dogs that have been fed unskinned carcasses. The animals have painful vesicles and pustules around the mouth, nostrils and ears. It is usually transmitted to the farmer or veterinary surgeon when working with the animal, especially at lambing time. After an incubation period of three to six days a single red, painful papule appears on the hand or forearm. It goes on to form a pustule and may become infected with other common bacteria with the development of an abscess or a spreading red, painful area (erysipelas).

RABIES (HYDROPHOBIA)

This is caused by a rhabdovirus which can infect all mammals. Dogs are the most common vector for human infection. The animals may become paralytic or aggressive and salivating. Man usually acquires it by a bite from an infected animal or occasionally by inhaling air in caves contaminated by infected secretions from vampire bats. The incubation period depends upon the closeness of the bite to the head, and it may be anything from ten days to one year. The disease produces mental changes with headaches, leading to spasm of the swallowing muscles causing hydrophobia or fear of drinking. These symptoms can usually be relieved with sedation but despite intensive care, paralysis and death ultimately result.

Pre-exposure immunisation for those most at risk and immediate post-exposure immunisation for anyone who has been bitten reduce the chance of rabies developing.

SWINE VESICULAR FEVER

This is caused by one of the picornaviruses. Pigs, when infected, develop vesicles around the hooves, snout and mouth. It is acquired by man by contact with infected secretions. The incubation period is uncertain. It may cause a prolonged, mild, influenza-like illness.

VESICULAR STOMATITIS

This is caused by one of the rhabdoviruses. It usually infects horses, cattle and pigs and results in the development of vesicles on the feet and mouth. Man is infected by contact with infected secretions. After a few days' incubation period, an influenza-like illness develops with some vesicles around the mouth.

Bacterial Infection

ANTHRAX

This is caused by *Bacillus anthracis*. Cattle, sheep and goats may be infected and can develop fever with convulsions and a throat infection (especially pigs), or they may simply be found dead. Humans are infected by contact with the infected animals, their carcasses or spores in the soil. After an incubation period of one to ten days, one of three types of anthrax may develop:

1. Cutaneous A pimple develops, becomes a blister and ultimately a black scab. Local swelling develops as does lymph node enlargement and septicaemia.

2. Pulmonary Pneumonia develops after the inhalation of spores.

3. Intestinal Bloody diarrhoea develops.

BORDETELLA

The common human infection whooping cough is caused by *B. pertussis*. Zoonotic infection is caused by *B. bronchiseptica*, and it is found in rabbits, dogs, cats and other animals. There is an incubation period of seven to ten days followed by bronchitis with cough and sputum. Endocarditis or a whooping cough-like illness may occasionally occur.

BRUCELLOSIS

Brucella abortus and *B. melitensis* are the most important species as far as man is concerned though many other *Brucella* also exist. *B. abortus* causes abortion in cattle, sheep and goats; orchitis in male animals; and infection in horses which may lead to chronic suppuration (discharge of pus) from the withers. It is transmitted to man by infected dairy products, contact with infected animals or the products of abortion. Inhalation of germs may also take place.

After an incubation period of one to six weeks, acute brucellosis symptoms may appear. These include fever with chills, sweating and weakness, a rash and an enlarged liver and spleen which may be painful. The symptoms of chronic brucellosis may be less well defined, with weariness, influenza-like aches and low back pain. Localised infection may develop in any organ, including the bones, brain, liver, lungs and heart. In view of the general lassitude, some patients are misdiagnosed as suffering from psychological disease.

CAMPYLOBACTER

This is caused by *Campylobacter fetus jejuni, C. fetus fetus* or *C. coli*. It is carried in the intestines of most animals and birds. The bacteria can cause abortion in cattle and sheep and may result in diarrhoea. Man is infected by eating undercooked meat or chicken and may be directly infected by dogs. After an incubation period of one to seven days, humans develop abdominal pain and diarrhoea with fever and influenza-like symptoms.

CAT SCRATCH FEVER

The organism that causes this is not known and there is no known disease in cats. Man is infected by cat bites or scratches although infection may result from scratches with other sharp objects possibly infected by the cat. After an incubation period of three to fourteen days, a red spot or papule develops with localised lymph node enlargement. There may be conjunctivitis (red eye) and influenza-like symptoms.

CORYNEBACTERIACEAE

It is a member of this group of bacteria that causes human-to-human spread of **diphtheria**. The zoonotic infections are caused by *Corynebacterium equi* affecting horses, cattle and pigs; *C. pseudotuberculosis (syn. C. ovis)* which infects sheep, cattle, horses, goats and deer; and *C. ulcerans* which infects cattle and horses. In the animals there may be abscesses and local lymph node swelling.

The organisms are transmitted to man in milk or by contamination from infected animal tissue. The incubation period lasts two to five days. *C. ulcerans* tends to cause a sore throat and occasionally a diphtheria-like illness. The others cause local infection and suppuration, with swollen lymph nodes. Occasionally septicaemia develops.

DF-2

This is important as an example of a newly discovered organism that can cause zoonotic infection. It has been discovered in the mouths of dogs, cats and other animals. It does not appear to be very infectious to healthy people but there is a risk for those who are more susceptible to disease. Thus this bacterium has been described in patients who have, for example, had their spleen removed, who are alcoholics or who have chronic respiratory disease. It is most commonly acquired by a dog bite (the area of the bite becoming ulcerated) or by the eye getting licked. Septicaemia can result as can intra-vascular blood clotting.

ERYSIPELAS

This is caused by *Erysipelothrix rhusiopathiae*, which may be present in pigs, fish and the soil. Pigs have a fever and a diamond-shaped skin rash, and they may develop arthritis and heart disease. It is transmitted to man directly, especially if the skin is abraded or wet. After an incubation period of two to seven days a painful, spreading, red skin rash develops. Arthritis may result, as may septicaemia.

GLANDERS

This is caused by *Pseudomonas mallei*. Horses and donkeys are the main reservoirs of this infection but dogs and cats may also become infected by eating infected raw meat. These animals develop a temperature with lung and nasal abscesses. Nodules on the skin may suppurate. Humans become infected from touching infected areas. After an incubation period of one to fourteen days a painful, inflamed ulcer develops which may slough off, and foul-smelling lung and nasal abscesses may develop.

LEPTOSPIROSIS

This is caused by one of the *Leptospira interrogans* organisms. *L. i. icterohaemorrhagiae* is caught from rats, *L. i. hardjo* infects cattle, *L. i. canicola* infects dogs, *L. i. bratislava* infects pigs and *L. i. ballum* may infect horses. No symptoms may be visible in these animals although there can be a range of illnesses similar to those in man. In cattle there is the additional feature of a rapid decline in milk yield.

Man becomes infected from the organisms present in urine, infected milk or water. These penetrate skin abrasions or the mucous membranes. The incubation period is from three to twenty days. Infection with *L. i. icterohaemorrhagiae* tends to be the most severe while that of the other varieties may only result in influenza-like illnesses. The classic clinical disease from *L. i. icterohaemorrhagiae* is Weil's disease in which there is jaundice, kidney disease and meningitis. Skin rashes, psychological changes, severe influenza-like symptoms and changes in many other organs of the body may occur.

LISTERIOSIS

This is caused by *Listeria monocytogenes*. It is widely distributed in animals (including man) and is present in the faeces of infected animals. It survives in the soil and, unlike most bacteria that infect man, continues to grow at low temperatures. Animals that contract the disease as opposed to carriers may have symptoms similar to those found in humans.

In view of the wide distribution of the organism the source of human infection is unclear, but it is probably acquired from contaminated food and milk. Because of its ability to multiply at low temperatures, the bacterium continues to grow in partially refrigerated food. This may give a false sense of security since refrigeration will not afford the protection given against other bacteria. Silage may also be heavily contaminated.

In man the incubation period is uncertain. The bacterium causes the development of multiple abscesses in the tissues, and vets or farmers may get skin abscesses. In pregnant women abortion may result, while in young babies, meningitis and pneumonia may occur. There may be a general, influenza-like illness with pneumonia or meningitis.

LYME DISEASE

This is caused by *Borrelia burgdorferi*. It is present in rodents and wild deer but may also infect dogs. Fever and arthritis may be present in some animals. It is transmitted to man by tick bites. After an incubation period of several days to weeks, there are three main presentations:

1. Skin rash This is a chronic, enlarging, red ring (*Erythema migrans*).

71

2. Arthritis   This may become chronic.

3. Neurological symptoms   There may be paralysis of individual nerves or chronic meningitis.

If severe these localised manifestations will be accompanied by influenza-like symptoms, lymph node enlargement, abdominal pain and occasionally cardiac disease.

MELIOIDOSIS

This is caused by *Pseudomonas pseudomallei*. It is found in pigs, sheep, goats and the soil. Animals lose weight due to the development of lung abscesses, and bone and joint infection. It is transmitted to man by direct infection of the skin, especially if the latter is softened by water. It may also be inhaled in dust. After a variable incubation period, subcutaneous or lung abscesses may develop.

PASTEURELLA MULTOCIDA

All animals and birds can be infected with this organism. They may not have symptoms though some develop severe haemorrhagic lung disease. Humans tend to become infected by a cat or dog scratch or bite. After an incubation period of forty-eight hours, inflammation develops around the affected area. The wound is slow to heal and may proceed to local abscess formation. Occasionally these bacteria cause lung abscesses and pus in the pleural cavity (between the lung and chest wall), especially in patients already suffering from chronic bronchitis.

BUBONIC PLAGUE

This is caused by *Yersinia pestis*, and it is found in the brown rat (*Rattus norvegicus*). The rat develops an illness similar to that in man unless it has acquired immunity, and the disease is transmitted from rat to man by fleas. In man the incubation period is two to six days. Initially the lymph nodes in such areas as the armpit and groin enlarge and may contain pus. This is followed by the development of severe pneumonia and the appearance of haemorrhages in the skin. It is these haemorrhages that give the illness the name the Black Death.

RAT BITE FEVER

This is caused by either *Streptobacillus moniliformis* or *Spirillum minus*. Rats infected with these bacteria are not usually diseased but excrete the infection in their urine and saliva. Man acquires the infection either by rat bites or from contaminated milk and water. The incubation period for *Streptobacillus* infection is three to ten days, while that for *Spirillum* infection is five to thirty days. In humans *Streptobacillus* causes a rash which may be haemorrhagic with influenza-like symptoms, joint pains and occasionally more general disease. *Spirillum* infection usually causes a local skin ulcer with enlarged lymph nodes and influenza-like symptoms. Occasionally more general infection results.

SALMONELLA INFECTION

There are many *Salmonella* bacteria. Infection can be passed between humans but it is most commonly acquired from eating undercooked infected poultry or other meat. The bacteria may, however, be acquired by direct passage from infected chickens, cows, sheep and goats. Eggs may also be contaminated. Diarrhoea is the most common symptom in animals.

In man there are two types of infection:

1. Food poisoning   After twelve to seventy-two hours, abdominal cramps and watery diarrhoea develop. This is frequently the result of infection with *S. enteriditis*, *S. typhimurium*, *S. pullorum* and *S. dublin*. This condition is relatively common.

2. Typhoid fever   This is caused by *S. typhi* which is invariably from a human source either by direct food contamination or watery infection. It can, however, be caused by *S. dublin*. After an incubation period of one to three weeks, headache, constipation or diarrhoea, abdominal pain, and general aches and pains develop. Rose-coloured spots may also appear on the skin. There are many possible complications due to localisation of infection in different organs of the body. In the Western world there is still the occasional case of typhoid fever but rarely a large outbreak. In countries with poor water supplies and sewerage systems, however, epidemics are more frequent.

STREPTOCOCCAL INFECTION

This is a common group of bacteria causing disease by transmission from human to human, but there are also some zoonotic infections. *Streptococcus suis* can infect the nose and throat of pigs with subsequent spread to the brain causing meningitis. *Streptococcus zooepidemicus* causes mastitis in cows and can also spread to other parts of the body.

The infections are spread to man by contact with lesions or infected meat. In man *S. suis* causes a fever and sometimes meningitis, while *S. zooepidemicus* causes a sore throat, pneumonia and occasionally endocarditis (infection of the lining of the heart) and nephritis (kidney disease).

TETANUS

This is caused by *Clostridium tetani*. This anaerobic organism (i.e. it grows without oxygen) is present in the faeces of domestic animals and can also survive in the soil. Humans are infected via a break in the skin, which may be no larger than a thorn prick in 20 per cent of patients. The spores germinate, and the growing organisms produce toxins which affect the nerves, resulting in muscle spasm and rigidity. If this affects the jaw and throat muscles, then lockjaw is said to be present. This is frequently followed by the development of respiratory disease including pneumonia.

Tetanus is prevented by immunisation. Farmers and their families should all be immunised and should maintain their level of protection by getting booster injections every few years. If an unprotected person is wounded, then immediate immunisation usually works. An antibiotic may also be given to kill the germs. This emergency approach is inconvenient and it may not be considered necessary to treat a small wound, a judgement that may have fatal consequences. If tetanus develops, the patient should be taken to hospital immediately to enable muscle relaxants and other intensive care to be given. In some countries anti-tetanus serum, which is produced in horses, is given although there is the risk of developing an allergic reaction to it.

Gas gangrene caused by *C. perfringens* may develop by the same mechanism.

TUBERCULOSIS

The human form of this disease is caused by *Mycobacterium tuberculosis hominus,* the bovine form by *M. bovis,* the bird form by *M. avium* and the fish form by *M. marinum.* Infected cattle may be asymptomatic or they may develop weight loss, bronchopneumonia, mastitis or disease in any part of the body. Humans get zoonotic tuberculosis by drinking infected milk and occasionally by contact with infected animals or their carcasses. The incubation period for the disease is variable since it may take several years after acquiring the infectious organisms for sufficient tissue damage to cause symptoms.

In man, bovine tuberculosis commonly causes enlarged lymph nodes in the neck and abdominal pain due to intestinal involvement. It may, however, infect any organ, including the lungs as with the human variety. *M. avium* infection tends to cause lymph node enlargement but may also cause pulmonary disease, especially if the lungs have previously been damaged by another disease. *M. marinum* causes skin ulcers.

TULARAEMIA

This is caused by *Francisella tularensis.* Its natural reservoir is in rabbits, hares and rodents, but it may infect other animals, especially if there is a heavy tick infestation. Infected animals may have a high temperature, become prostrate and may die. Ewes may abort and dogs develop soft swellings under the skin. Man is infected by the bite of a tick from an infected animal. After an incubation period of one to ten days there is usually an inflamed papule or ulcer which may suppurate. A high temperature, pharyngitis and pneumonia may result.

YERSINIOSIS

This is caused by *Yersinia pseudotuberculosis* and *Y. enterocolitica.* Many animals and birds are infected and they may develop diarrhoea, pneumonia or abortion. In rams the testes may also be affected. Man acquires the infection from unpasteurised milk although there may be other means of transmission. After an incubation period of three to seven days, abdominal pain and diarrhoea develop. This may be associated with a painful red, and later bruiselike, rash on the lower legs (erythema nodosum), pharyngitis and arthritis may also occur.

75

Rickettsial Infection

The *Rickettsia* are small, intra-cellular parasitic organisms similar in size to bacteria. They are chiefly transmitted to man from rodents by arthropods (ticks, mites, fleas and body lice). They cause Rocky Mountain spotted fever, Queensland tick typhus, rickettsialpox and typhus. They also cause Q fever.

Q FEVER

This is caused by *Coxiella burnetii* which occurs in sheep, cattle and goats, in which it may cause abortion. It is transmitted to man by ticks or by inhalation from dried aborted or infected tissue. The incubation period is two to four weeks. Unlike the other rickettsial diseases there is no rash. There is an influenza-like illness with cough and atypical pneumonia. The liver and heart valves may also be affected.

Chlamydial Infection

The chlamydial organisms are bacteria which grow only inside the living cells of the body. *Chlamydia trachomatis* causes a form of venereal disease, but more importantly it is responsible for eye infection in some countries. Although it is not a zoonotic illness in the usual sense of animal to human transmission, it may be transmitted from human to human by house flies. The main zoonotic disease caused by these organisms is psittacosis.

PSITTACOSIS

This disease is caused by *C. psittaci*, which infects parrots, ducks, turkeys, chickens and pigeons. It also infects sheep. Most infections are subclinical (i.e. the patient is infected but has no symptoms), but respiratory symptoms and abortion in sheep may occur.

It is transmitted to man by aerosols of infected material or dust from faeces, nasal discharge or the products of gestation and abortion. More recently, man-to-man transmission with *C. pneumoniae* has been described. The incubation period is four to fifteen days and there is usually an influenza-like illness. This may progress to atypical pneumonia and hepatic disease. The sheep variety may cause abortion in pregnant women.

Fungal or Mycotic Infection

Many fungi can infect man, and they occur on and in animals, in the soil and air and on vegetation. Varieties differ around the globe. The diseases caused are known as mycotic infections, and in this section, ringworm and histoplasmosis are dealt with in more detail than the other infections.

RINGWORM

Zoonotic ringworm is caused by the fungi *Trichophyton mentagrophytes* and *T. verrucosum*, which infect cattle and horses, and *Microsporum canis*, which infects dogs and cats. They cause grey patches on the animal's skin.

Man is infected by contact with the animal, its tackle or surroundings contaminated by the spores of the fungus. The incubation period is four to fourteen days. *T. verrucosum* causes an intensely itchy, red, pustular rash with subsequent scaling of the skin. It can also cause patchy hair loss and a red, scaly rash. The rash may have a ring-like appearance, hence the name ringworm.

HISTOPLASMOSIS

This is caused by *Histoplasma capsulatum*. In Africa, America and the Far East, birds and bats which are usually asymptomatic contaminate soil and houses with fungal spores from their droppings. Man gets the infection by inhaling the spores in dry dust. After an incubation of five to eighteen days, there is an influenza-like illness with subsequent chest discomfort and shortness of breath. Chronic disease of the lungs and other organs in the body may appear.

OTHER FUNGAL OR MYCOTIC INFECTIONS

These infections are common in farmers because of their contact with the soil and vegetation.

Coccidioidomycosis is caused by *Coccidioides immitis* in arid parts of America and is distributed in dust from uncultivated ground. It causes influenza-like symptoms, skin and lung infection and may cause disease in any organ.

Blastomycosis is caused by *Blastomyces dermatitidis*. It was thought to be confined to North America but has now been found in other American and in African countries. It is present in the soil, and dogs and horses are very susceptible to it. The infection can affect all parts of the body.

Paracoccidioidomycosis is caused by *Paracoccidioides brasiliensis* and is found in the soil and on vegetables in South America. It affects the lungs and many other parts of the body.

Cryptococcosis is caused by *Cryptococcus neoformans* in Europe and America. It is present in the soil and pigeons carry it on their beaks and in their droppings. It infects the lungs, brain and other organs.

Sporotrichosis is caused by *Sporothrix schenckii* and is found in America and Europe, where it is present on vegetation and timber. It spreads under the skin and occasionally affects the lungs.

Mycetoma This is a mixed infection caused by fungi, actinomycetes and bacteria. It affects the outdoor workers in semi-tropical countries, causing lumps and abscesses in the skin and bones.

Zoonotic Parasites

These are divided into the protozoal parasites and the many types of worm.

PROTOZOA

Babesiosis This is caused by the protozoa *Babesia* species. *B. bovis*, *B. bigemina* and *B. divergens* infect cattle. *B. caballi* infects horses, *B. canis* infects dogs and *B. microti* infects rats. The illness is usually mild in animals but may progress as in the human disease.

It is transmitted to humans by ticks, but clinical disease is rare unless the patient has no spleen. After an incubation period of one to twelve months the patient develops a malarial-like illness with severe chills followed by anaemia, jaundice and sometimes kidney failure.

Toxoplasmosis   This is caused by the protozoa *Toxoplasma gondii*. It infects many animals, but as far as man is concerned, the cat is the most important source of infection. Animals do not usually show signs of the infection, although sheep may abort in the final month of pregnancy.

Man is usually infected by cat faeces but may also acquire it from undercooked meat and goat's milk. A pregnant woman should avoid lambing ewes in view of the risk of infection to both herself and her foetus. During the first three months of pregnancy the foetus is less susceptible to the disease, but if infection does occur, abortion or stillbirth results. If the infection is at a later stage in pregnancy, the baby may be born with severe brain, eye, liver or splenic disease. Milder forms of brain or eye disease may only become apparent in later life.

In adults the disease is similar to glandular fever with a temperature, aches and pains, swollen glands and spleen and a general feeling of fatigue.

Cryptosporidiosis   This is caused by *Cryptosporidium muris* in mammals and *C. meleagridis* in birds. Man probably is infected by drinking water or milk contaminated by calves, mice, kittens, puppies and goats. The disease causes diarrhoea and abdominal cramps, especially in children.

Trypanosomiasis

There are two kinds. African is caused by the protozoa *Trypanosoma gambiense* and *T. rhodesiense*. Man is the main reservoir for this infection but it may be acquired from cattle in whom the disease is usually sub-clinical. It is transmitted by the tsetse fly. After an incubation period of three to twenty-one days a local lesion may appear, followed by generalised symptoms and then sleeping sickness with mental deterioration, sleepiness and subsequent convulsions.

South American is caused by *T. cruzi*. Cattle, dogs and cats are the main source of this infection and often have few symptoms although acute fever with lymph node and liver enlargement may result. There may also be heart disease. It is transmitted to man by blood-sucking triatomids (bugs) whose faeces may be rubbed into scratches or the eye. Five to fourteen days after the bite a local ulcer may develop with fever, enlarged lymph nodes, liver and spleen. Chronic disease of the heart and bowel may develop.

Leishmaniasis   There are many *Leishmania* species which can cause this disease. These infect wild animals and dogs outside Europe and are spread by the sand fly. Humans are the main reservoirs in some countries. Several forms of the disease exist, and symptoms are similar in both humans and animals.

There may be a skin nodule or ulcer and such lesions may affect the mouth. The generalised disease is a fever with an enlarged spleen and anaemia. Secondary infection with other bacteria may develop due to suppression of the immune system. The incubation period for the skin form of the disease is from one week to several months, while that for the generalised form is from one week to several years.

WORMS

Roundworms (Nematodes)

Trichinosis   This is caused by the roundworm *Trichinella spiralis*. The natural hosts are rats, carnivores and pigs, the latter being the animal responsible for human disease. The disease in pigs may be similar to that in man or it may be asymptomatic. Man is infected by eating undercooked pork, and the incubation period is from ten to fourteen days. Although there may be initial gastro-intestinal symptoms, the main problem arises from the passage of the worms into the tissues of the body, where they die and become bone-like. Pain and swelling of the tissues result.

Toxocariasis   This is caused by the roundworm *Toxocara canis* and *T. cati*. These infect dogs and cats, the former being the more important source of infection for humans. The worms cause few symptoms in the animals, unless there is a massive infection in the young. Humans are infected by hand or food contamination from dog faeces. The incubation period varies from weeks to months and is often subclinical. Older children tend to develop damage to the retina of the eye with resulting poor vision, while young children develop general disease due to migration of the worm larvae (young worms) through the tissues. There may be intestinal, lung and brain disease.

There are a number of other roundworm infections to which farmers and their families are more susceptible because of their contact with the soil and water. Strictly speaking, however, they are not all zoonotic infections:

Angiostrongyliasis   This is caused by the rat lungworm *Angiostrongylus cantonensis* in the Pacific Islands. It passes through snails and slugs to infect lettuce, crabs and prawns. It causes meningitis and brain disease in man.

Ascariasis   *Ascaris lumbricoides* is the largest roundworm to infect man and is more common in warm climates. Soil becomes contaminated by eggs in infected human faeces. Children are frequently infected by ingestion of the soil while adults are more frequently infected by eating unwashed vegetables. It causes lung and abdominal disease together with malnutrition.

Cutaneous larva migrans (creeping eruption)   This occurs in warm climates and is caused by *Ancylostoma brasiliense*, the cat and dog hookworm. The larva in soil contaminated with infected faeces enters the skin and causes an itchy, creeping skin rash.

Hookworms   *Ancylostoma duodenale* and *Nectator americanus* are passed from man to man by larvae in the soil in warm countries. They cause a skin rash, lung disease and severe anaemia.

Strongyloidiasis   *Strongyloides stercoralis*, found in warm countries, is passed directly or through the soil from man to man. It causes a rash, lung disease, abdominal pain and malnutrition.

Cestode Worms

Hydatid cysts   These are caused by *Echinococcus granulosus*. Sheep are the main reservoir for the worms but dogs become infected after eating raw offal. Sheep and dogs are usually asymptomatic. Man gets the disease by contact with the eggs in faeces from an infected dog. The incubation period is from months to years and the illness is caused by the growth of cysts in tissues such as the liver, brain or lungs.

Tapeworms   These are the cestode worms *Taenia saginata* affecting cattle and *T. solium* affecting pigs. The animals usually are without symptoms but may develop muscle pain or brain disease. Man is infected by eating undercooked meat, and the cysts develop into worms in the intestine. The incubation period lasts from one to two weeks during which the worms grow, with resulting nausea, weight loss and abdominal pain. Anaemia is present in more chronic infection. *Diphyliobothrium latum*, the fish tapeworm, produces a similar picture.

If human faecal eggs from *T. solium* are ingested, the larvae migrate throughout the body causing a more generalised infection known as **cysticercosis**. In this there is fever and generalised pain in the muscles. There may also be brain disease and skin nodules.

Trematode Worms

Liver flukes *Fasciola hepatica* and *F. gigantica* infect domestic and wild herbivores which may suffer from weight loss, anaemia and jaundice. The worms in the bile ducts of the liver lay their eggs, which pass into the faeces. On damp pasture, a miracidium hatches and enters a water snail. A cercaria leaves the snail and forms a cyst on water plants, and man is infected by eating these contaminated water plants. The worms hatch and migrate to the bile ducts of the liver. After a variable incubation period, fever, abdominal pain and jaundice result. The larvae may also cause a skin rash during their migration.

Schistosomiasis (bilharziasis) This is caused by *Schistosoma mansoni, S. haematobium* and *S. japonicum*. It is not a true zoonotic disease in that both the sufferer and the reservoir are man, although *S. japonicum* may infect cattle, horses, dogs and other animals. The intermediate stage of a water snail is required for the miracidium to mature to the cercarial stage which penetrates man's skin. This disease is included under zoonoses because due to their close contact with damp soil and water, farmers are likely to become infected in those countries where the disease is endemic.

The initial symptoms are those of an itchy skin rash. Some two to nine weeks later acute symptoms may develop, especially with *S. japonicum* which produces influenzal symptoms and diarrhoea. The liver and spleen may become enlarged and severe liver and bladder disease may develop.

Clonorchiasis Caused by *Clonorchis sinensis*, this infects fresh water fish, carp and salmon in Far Eastern countries. After ingestion man develops abdominal pain and liver disease.

Paragonimiasis This is caused by *Paragonimus westermani* and infects cats, dogs, cattle, foxes and pigs in warm Far Eastern countries. Crayfish and crabs become infected from these animals and they in turn infect man. Acute lung symptoms and abdominal pain result.

Prevention of Zoonotic Disease

All animals, including man, carry many bacteria and other infectious organisms on their skin, mouth, intestinal tract and sexual organs. It is therefore impossible to avoid contact with these during normal life and, indeed, it may not be desirable to avoid contact, since these organisms probably have a function. This can be seen when the normal body bacteria are killed with antibiotics, allowing more serious germs to infect the internal and external surfaces of the body, often with disastrous consequences.

It is important, therefore, to reach a correct compromise that allows as much freedom of contact with animals as possible with the minimum risk and the greatest cost effectiveness. That risk may depend upon the age of the farmer or his family. Thus, although toxoplasmosis may not be a serious disease in adults, it may cause death to the farmer's unborn baby. It may seem expensive to immunise cattle against some diseases or to treat them with anti-parasitic drugs. This cost has to be balanced against the risk of losing days off work during such times as sowing or harvesting. Often the decision has to be made by the society as a whole, by means of government departments and committees on health and safety, rather than by individual farmers. For example, intervention by the government, allied with cooperation from farmers, in the policy of slaughtering infected cattle has greatly reduced or abolished brucellosis in Northern Ireland.

The general principle of commonsense hygiene should always be applied. All animals which are obviously ill should be considered a health risk to the farmer and his family. Until veterinary advice is taken concerning the nature of the illness in an animal, there should be minimal skin contact unless protective clothing and gloves are worn.

SICK ANIMALS

Isolate sick animals.

Prevent disease spread to others.

Protect yourself and your family at the same time.

In some circumstances it may be necessary to wear government approved masks to avoid the inhalation of infected dusts (see Appendix 3). Hands and any other contaminated parts of the body must be washed in clean water, especially before meals.

Water is a potential threat in several ways. It may be contaminated by infectious agents, e.g. leptospiral organisms. Even clean water may be a problem, because if the skin becomes soaked and softened, this may allow many of the germs described in this chapter easier entry into the body. It is thus important to reduce the period of contact with water to a minimum and to use a conditioning or barrier cream before starting work. Another possible risk is the use of high pressure hoses to clean milking parlours and other areas, as the spray created may contain germs. It may be prudent to use a mask during this operation.

Intensive farming methods increase the frequency of contact between animals and hence the likelihood of zoonotic infection because of the greater likelihood of the transmission of germs between animals. There is also increased indoor contact with the farmer and thus the risk of zoonotic infection increases. If infection does occur in an animal and part of this animal is used to increase the protein content of food for other stock, an explosive increase in transmission may occur. Reputable livestock food producers will obviously try to avoid such an accident, but a small risk will always remain, particularly if a currently unrecognised pathogen appears. The farmer is at the same risk as other members of the community from contamination of modern processed, refrigerated and microwaved food. The use of refrigerated rather than fresh food in hot climates with unreliable sources of electricity is particularly dangerous.

Animals that have aborted and the aborted products are especially dangerous to women and the family of the farmer. It is essential that this risk is recognised and care taken.

Aborted products are a special infectious risk.

Farmers should make themselves aware of the incidence of various zoonotic diseases in their neighbourhood or country since there is such wide variation in the types of diseases and their incidence from region to region. For instance, in Northern Ireland 71 per cent of farmers of all ages had evidence of toxoplasma infection compared with only 20 per cent of other young people in the community and 60 per cent of older people. Thirteen per cent of dairy farmers had evidence of leptospiral infection as opposed to only

3 per cent of mixed farmers, and 26 per cent of farmers had evidence of Q fever in the past. Psittacosis had been a cause of infection in 11 per cent of farmers although only 5 per cent of the general population had had this illness. Awareness of these and similar figures should bring home the necessity for taking care on the farm.

Zoonotic diseases that can be acquired from different animals

FROM BIRDS
Newcastle disease
salmonella food poisoning
tuberculosis
psittacosis
cryptococcosis

St Louis encephalitis
campylobacter jejuni
yersinia pseudotuberculosis
tularaemia

FROM CATS
toxoplasmosis
cat scratch fever
cowpox (rare)

ringworm
pasteurella multocida
cryptosporidiosis

FROM CATTLE
cowpox
hand, foot and mouth
streptococcal infection
typhoid fever
anthrax
tuberculosis
Q fever
corynebacterium sp.
babesiosis
liver fluke

orf
vesicular stomatitis
salmonella food poisoning
campylobacter jejuni
brucellosis
leptospirosis
actinobacillus ligniersei
ringworm
tapeworms
cryptosporidiosis

FROM DEER
louping ill encephalitis
corynebacterium sp.

lyme disease

FROM DOGS
toxocara canis
rabies
bordetella bronchiseptica
lyme disease
babesiosis
cryptosporidiosis

hydatid cysts
orf (rare)
pasteurella multocida
ringworm
DF-2

Zoonotic diseases acquired from different animals (contd)

FROM FISH
tuberculosis
tapeworm
paragonimiasis

erysipeloid
clonorchiasis

FROM FOXES
rabies

FROM GOATS
orf
melioidosis
anthrax
corynebacterium sp.
cryptosporidiosis

Central European encephalitis
brucellosis
Q fever
toxoplasmosis

FROM HORSES
vesicular stomatitis
leptospirosis
ringworm

glanders
corynebacterium sp.
babesiosis

FROM PIGS
influenza (rare)
streptococcus suis
typhoid fever
yersinia pseudotuberculosis
leptospirosis
tapeworms

swine vesicular fever
salmonella food poisoning
melioidosis
erysipelas
trichinella worms
cysticercosis

FROM RODENTS
haemorrhagic nephritis
  (hantan fever)
lymphocytic choriomeningitis
yersinia pseudotuberculosis
leptospirosis (Weil's disease)

tularaemia
plague
rat bite fever
babesiosis
cryptosporidiosis

FROM SHEEP
orf (lambing season)
melioidosis
brucellosis
psittacosis
corynebacterium sp.
hydatid cysts

louping ill encephalitis
anthrax
Q fever
actinobacillus lignieresi
toxoplasmosis
liver fluke

Zoonotic diseases in hot climates or from the soil and vegetation

IN HOT CLIMATES
schistosomiasis
South American trypanosomiasis
leishmaniasis (skin, mouth,
  visceral)

trypanosomiasis
  (sleeping sickness)
clonarchiasis
paragoniasis

MAINLY IN THE SOIL AND VEGETATION
tetanus
coccidioidomycosis
paracoccidiodomycosis
sporotrichosis
hookworm
creeping eruption
angiostrongyliasis

listeriosis
blastomycosis
cryptococcosis
mycetoma
strongyloidiasis
ascariasis

Zoonotic diseases affecting various parts of the human body

BONES AND JOINTS
brucellosis
rat bite fever

lyme disease
tuberculosis

BRAIN, NERVES AND MUSCLE
rabies
encephalitis viruses
streptococcus suis
listeria
toxoplasmosis
trichinella spiralis
cysticercosis
leptospirosis

louping ill, BSE(?)
lymphocytic choriomeningitis
tetanus
actinobacillus lignieresi
sleeping sickness
hydatid cyst
lyme disease

EYES
Newcastle disease
toxocara canis

toxoplasmosis
cat scratch fever

GASTRO-INTESTINAL TRACT
salmonella food poisoning
campylobacter
hydatid cysts
schistosomiasis
tuberculosis
cryptosporidiosis

typhoid
yersinia pseudotuberculosis
tapeworms
liver flukes
leptospirosis

Zoonotic diseases affecting various parts of the body (contd)

INFLUENZA-LIKE ILLNESSES

REMEMBER: **Human influenza.**

Human respiratory viruses.

Farmer's lung-type illness.

Other non-zoonotic infections.

---

swine vesicular fever
hantanvirus
salmonella infections
Q fever
psittacosis
babesiosis
leishmaniasis
rat bite fever

vesicular stomatitis of
   cattle and horses
streptococcus suis
brucellosis
tularaemia
toxoplasmosis
tuberculosis
leptospirosis

KIDNEYS AND URINARY TRACT
hantanvirus
leptospirosis

schistosomiasis

LUNGS
bordetella bronchiseptica
melioidosis
anthrax
psittacosis
corynebacterium equi
tuberculosis

tularaemia
plague
Q fever
actinobacillus lignieresi
hydatid cyst

MOUTH AND THROAT
hand, foot and mouth disease
leishmaniasis

corynebacterium ulcerans

SKIN
cowpox
hand, foot and mouth
  disease
tularaemia
melioidosis

pseudocowpox
orf
vesicular stomatitis of cattle
  and horses
glanders

SKIN (contd)

yersinia pseudotuberculosis	plague
pasteurella multocida	erysipeloid
gas gangrene	anthrax
actinobacillus lignieresi	listeria
ringworm	corynebacteria
cysticercosis	leishmaniasis
tuberculosis	lyme disease
cat scratch fever	rat bite fever

Appendix 1
Some of the Prescribed Occupational Diseases in the UK that May Affect Farmers

Conditions Caused by Physical Agents

Bursitis of the knee (housemaid's knee) or elbow (tennis elbow)
Cellulitis of the hand (infection after injury)
Tenosynovitis (inflamed tendons of hand or forearm)
Occupational deafness
Episodic blanching (vibration white finger)
Inflammation or ulceration of the mouth or mucous membranes (by dusts, liquids or vapours)
Non-infective dermatitis or skin rash

Poisoning

Arsenic
Mercury
Methyl bromide
Di-nitro-ortho-cresol and dinitrophenol
Phosphorus or phosphorus compounds
Anti-cholinesterase inhibitors

Conditions Caused by Infections

Cellulitis (infection of any patch of skin)
Anthrax
Glanders
Brucellosis
Streptococcus suis
Leptospirosis
Ankylostomiasis

Tuberculosis
Chlamydiosis or psittacosis (bird or sheep)
Q fever

Diseases Due to Allergy or Immunology

Industrial asthma
Dermatitis
Extrinsic allergic alveolitis (e.g. farmer's lung, bird-fancier's lung, mushroom-grower's lung, etc.)

Appendix 2
Notifiable Infectious Diseases in the UK

In Humans

Acute encephalitis/meningitis: bacterial
Acute encephalitis/meningitis: viral
Meningococcal septicaemia
Anthrax
Chickenpox
Cholera
Diptheria
Dysentery
Food poisoning
Gastro-enteritis (persons under two years only)
Hepatitis A
Hepatitis B
Hepatitis unspecified: viral
Legionnaire's disease
Leptospirosis
Malaria
Measles
Mumps
Paratyphoid fever
Plague
Poliomyelitis: acute
Rabies
Relapsing fever
Rubella
Scarlet fever
Smallpox
Tetanus
Tuberculosis: pulmonary & non-pulmonary
Typhoid fever
Typhus
Viral haemorrhagic fevers
Whooping cough
Yellow fever

In Animals

African horse sickness
African swine fever
Anthrax
Aujeszky's disease
Bovine spongiform
  encephalopathy (BSE)
Brucella melitensis in cattle
Cattle plague in ruminants and
  swine
Classical swine fever
Contagious equine metritis
Dourine in horses, asses,
  mules and zebras
Enzootic bovine leukosis
Epizootic lymphangitis in
  horses, asses and mules
Equine encephalomyelitis
Equine infectious anaemia

Foot and mouth disease in
  ruminants and swine
Fowl pest (fowl plague,
  Newcastle disease and
  paramyxovirus) in poultry
  of all kinds
Glanders and farcy in horses,
  asses and mules
Paramyxovirus in pigeons
Pleuro-pneumonia in cattle
Rabies
Sheep pox
Sheep scab
Swine vesicular disease (SVD)
Warble fly
Teschen disease of pigs
Tuberculosis in cattle (certain
  forms)

Appendix 3
Face Masks and Respiratory Helmets

No face mask can guarantee complete protection from dust or fume inhalation. The general principle of **avoidance** rather than **protection** should therefore always be applied.

When selecting a mask or respiratory helmet it is important to remember several points:

1. **Always choose one which has a government approval stamp,** e.g. in the UK for dust such as mouldy hay, a mask providing protection to at least British Standard (BS) 6016 type B should be chosen. In Europe, new, more rigorous standards (TM 4 Part 9.1, prEN 140, prEN 141, prRN 143) will be introduced in 1992, and these will give better protection. For other dusts and toxic fumes check the legal requirements with a salesman or the Ministry of Agriculture before purchase. More stringent criteria for protection

are regularly assessed as new information becomes available.

2. Choose protection that can be worn comfortably for the length of time you will normally use it.   Remember that you will frequently be physically active during periods of use and this increases the volume of air you need to breathe. Because of resistance to the passage of air through some masks, especially as they remove the dust, breathing becomes difficult. The only way to ensure that the mask is really suitable for you is to try one out in the same way as when buying any other apparatus for the farm or home. Incorrectly used or old discarded masks will not protect.

Protectors are of three main types, and each has been given a comparative cost (ranging from 1 to 300 units) and protection factor (ranging from a low of 1 to 50):

Simple light masks   These consist of disposable moulded masks or a lightweight aluminium holder with a disposable gauze pad. Under high workloads, they may silt up, and frequent replacement may be required.

    Comparative cost: 1–2 units
    Approximate protection factor: 4

Cartridge respirators   These consist of a rubber-type fitted mask with disposable or removable cartridge filters. These masks are a little heavier, but the valve mechanisms reduce moisture and heat build-up. Although more frequently used for protection against gases, these can also be used for dusts and sprays.

    Comparative cost: mask 10–20 units
                                    filter 1–4 units
    Approximate protection factor: 50

Helmets   These cover the head and face. They have a battery-operated fan which draws air through filters and forces it across the face; for some jobs a clean air line from a compressed air cylinder is used. This has the advantage of less heat and moisture build-up but the disadvantage of cost, size and weight. Masks using external air supplies have additional costs but can give full protection.

    Comparative cost: helmet 200–300 units
                                    filter 3 units
    Approximate protection factor: 10

Index

Farming Press Books

Below is a sample of the wide range of agricultural and veterinary books published by Farming Press. For more information or for a free illustrated book list please contact:

Farming Press Books, 4 Friars Courtyard
30–32 Princes Street, Ipswich IP1 1RJ, United Kingdom
Telephone (0473) 241122

Practical Accounting for Farm & Rural Business ● BEN BROWN

Covers the full range of accounting needs from data collection through profit and loss to analysis of results.

Farming and the Countryside ● ERIC SOPER & MIKE CARTER

Traces the middle ground where farming and conservation meet in co-operation rather than confrontation.

The Veterinary Book for Sheep Farmers ● DAVID HENDERSON

A wide-ranging, detailed guide to the prevention of sheep ailments, increased lamb output and the diagnosis and treatment of disease.

A Veterinary Book for Dairy Farmers ● ROGER BLOWEY

Deals with the full range of cattle and calf ailments, with the emphasis on preventive medicine.

Organic Farming ● NICOLAS LAMPKIN

An outstanding wide-ranging account of the principles and practice for livestock and crops.

The TV Vet Horse Book ● EDDIE STRAITON

300 photographs and concise text give clear guidance to the recognition and treatment of horse ailments.

Farming Press Books is part of the Morgan-Grampian Farming Press Group which publishes a range of farming magazines: *Arable Farming, Dairy Farmer, Farming News, Pig Farming, What's New in Farming*. For a specimen copy of any of these please contact the address above.